PEDIATRICS, CHILD AND ADOLESCENT HEALTH

ENVIRONMENTAL HEALTH

HOME, SCHOOL AND COMMUNITY

PEDIATRICS, CHILD AND ADOLESCENT HEALTH

JOAV MERRICK –SERIES EDITOR –

NATIONAL INSTITUTE OF CHILD HEALTH AND HUMAN DEVELOPMENT, MINISTRY OF SOCIAL AFFAIRS, JERUSALEM, ISRAEL

Child and Adolescent Health Yearbook 2011
Joav Merrick (Editor)
2013. ISBN: 978-1-61942-782-2

Child and Adolescent Health Yearbook 2012
Joav Merrick (Editor)
2012. ISBN: 978-1-61942-788-4

Tropical Pediatrics: A Public Health Concern of International Proportions
Richard R Roach, Donald E Greydanus, Dilip R Patel, Douglas N Homnick and Joav Merrick (Editors)
2012. ISBN: 8-1-61942-831-7

Child Health and Human Development Yearbook 2011
Joav Merrick (Editor)
2012. ISBN: 978-1-61942-969-7

Child Health and Human Development Yearbook 2012
Joav Merrick (Editor)
2013. ISBN: 978-1-61942-978-9

Developmental Issues in Chinese Adolescents
Daniel TL Shek, Rachel CF Sun and Joav Merrick (Editors)
2012. ISBN: 978-1-62081-262-4

Positive Youth Development: Theory, Research and Application
Daniel TL Shek, Rachel CF Sun and Joav Merrick (Editors)
2012. ISBN: 978-1-62081-305-8

Understanding Autism Spectrum Disorder: Current Research Aspects
Ditza A Zachor and Joav Merrick (Editors)
2012. ISBN: 978-1-62081-353-9

Positive Youth Development: A New School Curriculum to Tackle Adolescent Developmental Issues
Hing Keung Ma, Daniel TL Shek and Joav Merrick (Editors)
2012. ISBN: 978-1-62081-384-3

Transition from Pediatric to Adult Medical Care
David Wood, John G Reiss, Maria E Ferris, Linda R Edwards and Joav Merrick (Editors)
2012. ISBN: 978-1-62081-409-3

Guidelines for the Healthy Integration of the Ill Child in the Educational System: Experience from Israel
Yosefa Isenberg
2013. ISBN: 978-1-62808-350-7

Chinese Adolescent Development: Economic Disadvantages, Parents and Intrapersonal Development
Daniel TL Shek, Rachel CF Sun and Joav Merrick (Editors)
2013. ISBN: 978-1-62618-622-4

University and College Students: Health and Development Issues for the Leaders of Tomorrow
Daniel TL Shek, Rachel CF Sun and Joav Merrick (Editors)
2013. ISBN: 978-1-62618-586-9

Adolescence and Behavior Issues in a Chinese Context
Daniel TL Shek, Rachel CF Sun and Joav Merrick (Editors)
2013. ISBN: 978-1-62618-614-9

Advances in Preterm Infant Research
Jing Sun, Nicholas Buys and Joav Merrick
2013. ISBN: 978-1-62618-696-5

Internet Addiction: A Public Health Concern in Adolescence
Artemis Tsitsika, Mari Janikian, Donald E Greydanus, Hatim A Omar and Joav Merrick (Editors)
2013. ISBN: 978-1-62618-925-6

Promotion of Holistic Development of Young People in Hong Kong
Daniel TL Shek, Tak Yan Lee and Joav Merrick (Editors)
2013. ISBN: 978-1-62808-019-3

Human Developmental Research: Experience from Research in Hong Kong
Daniel TL Shek, Cecilia Ma, Yu Lu and Joav Merrick (Editors)
2013. ISBN: 978-1-62808-166-4

Chronic Disease and Disability in Childhood
Joav Merrick
2013. ISBN: 978-1-62808-865-6

Break the Cycle of Environmental Health Disparities: Maternal and Child Health Aspects
Leslie Rubin and Joav Merrick (Editors)
2013. ISBN: 978-1-62948-107-4

Environmental Health Disparities in Children: Asthma, Obesity and Food
Leslie Rubin and Joav Merrick (Editors)
2013. ISBN: 978-1-62948-122-7

Environmental Health: Home, School and Community
Leslie Rubin and Joav Merrick (Editors)
2013. ISBN: 978-1-62948-155-5

Child Health and Human Development: Social, Economic and Environmental Factors
Leslie Rubin and Joav Merrick (Editors)
2013. ISBN: 978-1-62948-166-1

Children, Violence and Bullying: International Perspectives
Joav Merrick, Isack Kandel and Hatim A Omar (Editors)
2013. ISBN: 978-1-62948-342-9

Playing with Fire: Children, Adolescents and Firesetting
Hatim A Omar, Carrie Howell Bowling and Joav Merrick (Editors)
2013. ISBN: 978-1-62948-471-6

Pediatrics, Child and Adolescent Health

Environmental Health

Home, School and Community

I. Leslie Rubin, M.D.

and

Joav Merrick, M.D.

Editors

New York

For permission to use material from this book please contact us:
Telephone 631-231-7269; Fax 631-231-8175
Web Site: http://www.novapublishers.com

Library of Congress Cataloging-in-Publication Data

Library of Congress Control Number: 2013948784

ISBN: 978-1-62948-155-5

Published by Nova Science Publishers, Inc. † New York

Contents

Preface

Children living in circumstances of social and economic disadvantage are at high risk for experiencing health problems caused or exacerbated by environmental factors. They are not only vulnerable to adverse chemical and physical factors in their environment, but often become trapped in the cycle of environmental health disparities due to low parental health literacy, limited social capital, and a lack of access to comprehensive healthcare and appropriate educational services. They will only be able to escape that cycle when children's environmental health becomes a priority and the resources, energy and creativity are amassed and allocated to address environmental health concerns. In this book, we look at issues concerned with the home environment, the school and the community. Investing in children's environmental health will prove cost-beneficial in the long run.

Chapter I – Since 2005, Innovative Solutions for Disadvantage and Disability (ISDD) and the Southeast Pediatric Environmental Health Specialty Unit (SE PEHSU) at Emory University have held annual programs of student-driven research projects that focus on environmental health disparities of children vulnerable as a result of social and economic disadvantage. Each year, creative projects generated by an interdisciplinary group of university students focusing on children's environmental health disparities are selected for participation in the Break the Cycle Program for that year. The students are expected to present their projects at a national meeting and then write up their projects for publication.

The ISDD and SE PEHSU are pleased to publish this book to introduce the students' work to a wider audience, raise awareness of children's environmental health, and inspire additional research and intervention strategies to reduce chidlren's environmental health disparities.

Chapter II – Environmental health disparities (EHD) are disease issues that disproportionately affect particular populations due to complex interactions with environmental conditions. Perpetuation of EHD can be viewed as a closed "cycle" in which factors such as poor living conditions and limited health care exacerbate disease states. The purpose of this study was to investigate whether using an undergraduate curriculum that promotes awareness of EHD among students to develop a potential for leadership and empower them to consider a career and community involvement that could help reduce EHD and develop creative ways to "break the cycle". In the fall of 2007, sophomore and senior students in the biology department at Spelman College examined a case study of polychlorinated biphenyl (PCB) contamination on an African-American community in Anniston, Alabama, a landmark case of EHD. Students were interviewed following the conclusion of the course regarding their attitudes and awareness of EHD.We predicted three qualitative outcomes from our study: 1) students would show increased interest in health disparity careers, 2) students would show increased interest in non-medical environmental health disparities careers, 3) students would show increased awareness/interest in community involvement with environmental health disparities issues. Data from our interviews supported some of our predictions. Implications for future use of undergraduate curriculum to promote interest in the cycle of environmental health disparities are discussed.

Chapter III – Substandard housing conditions have been linked to widespread childhood environmental health ailments, including two of the leading causes of childhood morbidity: lead poisoning and asthma. In 2009, the US Surgeon General called for action around healthy homes. Improving home health environments can alleviate the cycle of childhood morbidity and mortality. The North Carolina Department of Environment and Natural Resources Children's Environmental Health Branch is working to build capacity at the state-level to expand the childhood lead poisoning prevention program to respond to additional in-home environmental health issues. To achieve this objective, North Carolina must consider recommendations for assessment, management, and evaluation. This paper will situate healthy homes on the national public health agenda; discuss ways that healthy homes programs address children's environmental health disparities; introduce the NC Healthy Homes Initiative; explore current healthy housing efforts in North Carolina through an examination of the Guilford County Healthy Homes Initiative; and provide recommendations for the NC Healthy Homes Initiative to address children's environmental health disparities.

Chapter IV – Grandparents Raising and Nurturing Dependents with Disabilities (Project GRANDD) is a support program for grandparents who are the primary caregivers of grandchildren with developmental disabilities, chronic illness, behavior or learning difficulties. A critical problem contributing to diminished quality of life for the family is the grandparents' own poor health, as well as lack of access to available resources. This project pairs nursing students from the local university with grandparents participating in the GRANDD support program to improve the caregiving grandparent's health. Methods: Twelve nurse practitioner students were paired with six grandparents. Students performed home visits and assessments, health histories, environmental assessments and windshield surveys over the course of several weeks. Follow-up visits for medication reconciliation, referral to resources, or for patient health education were conducted in person or via phone. Results: The grandparents demographic characteristics were: African American females, on fixed incomes on or near poverty level, had multiple co-morbidities, and showed reluctance in seeking help because of their role as primary caretaker. Nursing students demonstrated increased awareness of the cycle of disability and disadvantage, and improved skills that will enable them to aid vulnerable individuals in accessing community resources. Conclusions: This pilot program had a positive impact on students and grandparents. Much is needed regarding outcome measurement and standardization of methods, but the initial results demonstrate a beneficial relationship between nurse practitioner students and Project GRANDD participants, contributing to the ongoing effort to break the cycle of disadvantage and disability in this vulnerable population.

Chapter V – In this chapter the authors assessed the impact of health problems in children and young adults with spina bifida on their health-related quality of life. Design: They surveyed 124 children and young adults with spina bifida in central Florida identified through multiple sources including multi-specialty clinic rosters, Title V clinic registries, and community-based spina bifida associations. The Child Health Questionnaire, (CHQ) and SF-36 were used to assess health related quality of life (HRQOL). Results: The sample included 84 children aged 5 to 17 years old and 40 young adults age 18 to 30 years old. In the past year, more than half (55%) had experienced a bladder infection, 40% pressure sores, 39% an allergy to latex, 38% scoliosis or kyphosis, 32% chronic pain, and 27% a weight problem. Children with chronic pain (versus those without chronic pain) reported the largest difference in HRQOL scores in both the psychosocial or physical health areas: physical HRQOL: 24.7 vs. 37.5, $p < .01$; psychosocial HRQOL: 48.7 vs.41.4, $p < .01$.

Adults with chronic pain reported significantly lower psychosocial health scores (46.0 vs. 58.1, $p < .01$). Other conditions associated with significantly reduced HRQOL scores were pressure sores and bladder infections in children and latex allergy in adults. Conclusions: In both children and adults with spina bifida, chronic pain has a substantial negative impact on physical and psychosocial health-related quality of life. In order to maximize the quality of life of persons with spina bifida, increased efforts are warranted to identify and control chronic pain.

Chapter VI – Children with intellectual or behavioral disabilities have numerous challenges in their lives, including navigating their world without injury, developing appropriate interpersonal interactions, and calming of anxiety and fear. This case report is to raise awareness among an audience of individuals focused on the study, care, and management of persons with intellectual and developmental disabilities of the availability of service dogs to assist children with diagnoses including, but not limited to, fetal alcohol spectrum disorders (FASD) and autism. The authors are presenting findings from an informal observation of the first specifically trained service dog for the assistance of a child affected by fetal alcohol spectrum disorder (FASD). Method: They interviewed a family who has a child with FASD to collect information on the impact of the addition of the specially trained canine. Our target family relayed observations on the behavioral interactions at both the home and external environment of the child in the six months before receipt of the dog, at the time of receipt and training, and three months post incorporation into the family. The authors also sought information from various providers of service dogs on their uses of these assistance animals. Result: Perceived improvement in the family's ability to interact with their child, and the child's ability to interact with his environment. Conclusion: Service dogs have been shown to help individuals with a multitude of medical and mental health challenges. This case report suggests their use be considered for or offered to children with FASD, a spectrum consisting of multiple intellectual and behavioral disabilities. When an intervention like this is available to improve the quality of life of affected children and their families, the model needs to be replicated, and made available across socioeconomic lines.

Chapter VII – In times of heightened awareness regarding natural and man-made disaster threats, it is imperative that communities and families are prepared for catastrophic events. Research has shown schools are ill-prepared to manage many of these situations and parents are unaware of their children's schools disaster plans. This chapter was conducted to ascertain parents' current knowledge of school disaster preparedness. Methods: Investigator-administered surveys that addressed knowledge in areas of preparation, incident response, and recovery were given to parents of school-aged children who presented to a Pediatric Emergency Department (ED). Results: 340 parents were surveyed. 54% of parents felt that their children's school was not located at or near a potential disaster site. 60% knew of the existence of a disaster plan at the school and 80% did not know where their child would be in the event of an evacuation. 25% of parents did not know if emergency equipment was available at the school and 62% were not aware of the existence of a "crisis team". Over 36% of parents wanted to be involved in preparing or updating the schools' disaster plans. Conclusions: Even in the atmosphere of heightened disaster awareness, significant gaps exist in parental knowledge regarding their school's readiness for disaster. Improving parental awareness of the details of their school's disaster plan is necessary for adequate readiness for this very real threat.

Chapter VIII – Adolescents with special health care and educational needs face significant obstacles as they age out of traditional child health care and education. Transition services is the term used to describe a set of services and supports designed to assist adolescents in adjusting to the change from school to work, home to independence, and pediatric to adult health care. This project partnered with three Exceptional Student Education Centers (ESE) as part of the Duval County Public School system to study the current health and social needs and health care transition needs of adolescents ages 16 to 22 years with complex developmental and health care needs. The study included a survey of parents of adolescents with special health care needs, in-depth key informant interviews and focus groups. A majority reported that their youth has a doctor or nurse who knows them well, a quarter indicated mental and dental services were needed. With respect to transition services, 60% of parents said that their adolescent's health care provider discussed how health care needs would change as the youth approached adulthood. However, only 1/3 of parents had received information on obtaining health insurance or reported that a provider developed a plan to address changing needs. In-depth interviews and focus groups revealed barriers to care which included actual physical needs of the adolescents, cost of medications, difficult social environments and lack of coordination of services. Ideas for reforms of the system included campus health clinics and psychiatric consultation, and health care transition guides for parents.

Chapter IX – The role of physical school environment on student health and education is becoming better understood. There is a growing body of literature that indicates improved physical environments in schools (e.g. indoor air quality, lighting and acoustic conditions) can enhance student outcomes. In parallel, the green building movement is centered around designing buildings, including schools, that are more sustainable in order to decrease energy consumption, minimize environmental impact and create healthier spaces for occupants. This paper synthesizes findings from both green design studies and school outcomes studies to provide a systematic evaluation of the potential impacts of green school design features on student outcomes. Three inter-related topics are covered in detail: (I) overview of the "green" concept, including existing guidelines for "greening" schools, attitudes towards green schools and condition of the physical environments in non-green schools; (II) potential effects of the physical environment on school children, including documentation of national statistics and summary of findings from school research studies; (III) synthesis of findings, including a discussion of the knowledge gaps in the field of green school research and conclusions.

Chapter X – The benefits of screening potential school sites for environmental hazards and the special vulnerability of children to environmental pollutants and toxics are well understood. Less apparent, and less widely discussed until recently, are other health concerns associated with school location, such as ready access to healthy food options and opportunities for safe walking and bicycling. In recent years, a variety of stakeholders interested in children's wellness in school settings have expressed concerns about the trend to place larger schools further away from the communities they serve. In some cases, this trend can worsen environmental health risks, such as air pollution, and reduce opportunities for physical activity, with disproportionate impacts on communities already burdened with multiple stressors. This paper describes a project in Georgia that helps school system decision-makers more fully integrate children's environmental health and wellness concerns into all aspects of school siting planning, through a training curriculum based on the U.S. Environmental Protection Agency's 2011 School Siting Guidelines. Two training modules have been developed. Preliminary feedback indicates that the process itself—convening a diverse set of stakeholders with different expertise related to children's environmental health, community planning, and school system planning—results in new partnerships and the consideration of other perspectives. Better integration of municipal and school planning and improved understanding of environmental health considerations can improve public health outcomes for the entire community.

Chapter XI – This chapter describes a project using the"Media Madness/ Getting the Rap on the Media" model and a conceptual framework developed by the Jane Fonda Center. The purpose of the project was to provide an educational program to reduce the influence of the media on the sexual attitudes and behaviors of youth. The project and intervention is an enhancement of a media curriculum, focused on the subject of teen dating violence, self-worth, and the Cycle of Disadvantage and Disability (CDD) model. This project has three components: 1) Research on teen dating violence, 2) Exploration of the relationship between the CDD, self-worth, teen dating violence, and the media and 3) Educating, engaging, and helping adolescents to develop healthy relationships. This paper documents the first phase of the Media Refection Project: risk assessment, research, methodology, and intervention development. The next phase is implementation and evaluation.

Chapter XII – Research studies have explored the association between exposure to media violence and aggressive behaviors of youth, including relationship abuse. This project was an enhancement of a media curriculum that focuses on the subject of teen dating violence (TDV)/ intimate partner abuse (IPA), self-worth and the cycle of disadvantage (CDD) model. This chapter documents implementation and evaluation. The educational program (intervention) was pilot tested to evaluate feasibility and impact among the target group using two methods of program evaluation: process and impact. Methods: Twenty-one high school students in metropolitan Atlanta participated in a pilot test of the media intervention. The intervention group received the enhanced media curriculum over a two day period. At the end of the intervention, 8 of the 21 participants were selected to be a part of a focus group to measure if the evaluation questions were addressed and if the various objectives had been achieved. Results: The use of TDV/IPA case studies in the media was a useful approach for this intervention. The focus group identified emotional and verbal abuse as potential causes of physical abuse in a teen dating relationship. Impact evaluation suggests an increase in perceived risk of being in a TDV/IPA relationship among the focus group participants. However, some teens were already aware of media influences on aggressive behaviors and could identify media messages. Conclusions: The intervention highlights some positive outputs for project feasibility and impact. A pilot test evaluation was necessary but limited the assessment of reliability of the results.

Chapter XIII – Urban, low-income minority adolescents are disproportionately affected by numerous preventable morbidities, including poor school performance, obesity and teen pregnancy. Due to the complex interplay of social, cultural and economic factors that contribute to the morbidities, no single intervention has shown consistent success in categorically improving outcomes. Recent research suggests a common root problem is lack of empowerment. Effective positive youth development (PYD) initiatives can prevent many morbidities by fostering empowerment in at-risk youths. However, little is published on how to implement and sustain such interventions. Methods: Cyclopedia is an annual summer curriculum of mentored bicycle rides consisting of educational lessons, local exploration, and creative self-expression (photography and writing). It is structured on a PYD model and is supported by a core partnership between a medical institution and a community-based organization (CBO). Results: Cyclopedia has completed two full seasons with expansion from 18 boys to 50 boys and girls. The cost per trip has dropped from roughly $400 per rider to $100 per rider. The number of rides increased from 13 to 20, and total mileage increased from 740 to 870 miles per year. However, miles per rider dropped from 41 to 17, and writings per rider dropped from 2.6 to 0.5. Conclusions: Cyclopedia, supported by a core partnership between a medical institution and a CBO, shows significant promise in cost-effectively reaching large numbers of at-risk children with PYD programming. Ongoing curriculum adjustments aim to address decrements in individual participation.

In: Environmental Health
Editors: IL Rubin and J Merrick
ISBN: 978-1-62948-155-5

Chapter I

Break the cycle of environmental health disparities

***I Leslie Rubin, MD*[*1,2,3,4], *Robert J Geller, MD,*[2,5], *Janice Nodvin, BA*[1,2], *Michele Marcus, MPH, PhD*[2,6], *Maeve Howett, RN, PhD*[2,7], *Benjamin A Gitterman, MD* and *Joav Merrick, MD, MMedSci, DMSc*[9,10,11]**

[1]Innovative Solutions for Disadvantage and Disability, Atlanta, GA, US
[2]Southeast Pediatric Environmental Health Specialty Unit, Emory University, Atlanta, GA, US
[3]Department of Pediatrics, Morehouse School of Medicine, Atlanta, GA,US
[4]Developmental Pediatric Specialists, Atlanta, GA, US
[5]Georgia Poison Center, Grady Health System, Atlanta, GA, US
[6]Rollins School of Public Health at Emory University, Atlanta, GA, US
[7]Nell Hodgson Woodruff School of Nursing, Emory University, Atlanta, Georgia, US
[8]Departments of Pediatrics and Public Health, George Washington University Schools of Medicine and Public Health and Health Services, General and Community Pediatrics, Children's National Medical Center, Washington DC, US
[9]National Institute of Child Health and Human Development, Jerusalem, Israel
[10]Office of the Medical Director, Division for Intellectual and Developmental Disabilities, Ministry of Social Affairs, Jerusalem, Israel
[11]Kentucky Children's Hospital, University of Kentucky, Lexington, Kentucky, US

[*] Correspondence: I Leslie Rubin, MD, Research Associate Professor, Department of Pediatrics, Morehouse School of Medicine and President, Innovative Solutions for Disadvantage and Disability, 776 Windsor Parkway, Atlanta, GA 30342, United States. E-mail: lrubi01@emory.edu.

Introduction

Since 2005, Innovative Solutions for Disadvantage and Disability (ISDD) and the Southeast Pediatric Environmental Health Specialty Unit (SE PEHSU) at Emory University have held annual programs of student-driven research projects that focus on environmental health disparities of children vulnerable as a result of social and economic disadvantage. Each year, creative projects generated by an interdisciplinary group of university students focusing on children's environmental health disparities are selected for participation in the Break the Cycle Program for that year. The students are expected to present their projects at a national meeting and then write up their projects for publication.

The ISDD and SE PEHSU are pleased to publish this book to introduce the students' work to a wider audience, raise awareness of children's environmental health, and inspire additional research and intervention strategies to reduce chidlren's environmental health disparities.

The student projects in this book focus on social and economic disparities and their impact on the health and development of children in their homes, their schools and their communities, that is where the children live, learn and play. The designs and methodologies vary significantly and reflect a diversity of styles and disciplines. However, all underscore the need for active participation and intervention and further environmental health disparity research that can be used to inform public policy, public health initiatives, and the training of future leaders in a wide range of academic and professional disciplines.

Child health

A myriad of environmental factors interact in complex ways to affect children's health and well-being. Adverse environmental factors can cause, exacerbate, or complicate serious childhood health problems such as asthma, obesity, hypertension and diabetes, as well as neurodevelopmental conditions, such as learning disabilities and attention-deficit-hyperactivity disorder. Unfortunately, many of these environmental factors that operate through direct and indirect mechanisms are not clearly understood. We need to understand this better if we are to help children live longer, healthier, happier and more successful lives (1-3).

Efforts to promote children's health and reduce or eliminate childhood health problems must incorporate an understanding, not only of the how environmental factors, such as chemical toxins, cause damage to organ systems, but also how social and economic factors as well as physical spaces and structures influence the health, growth and development of children. Home and school environments, open spaces, and community infrastructure help shape children's diets and influence levels and types of physical activity. They also influence children's interactions with other people and the natural world around them. The condition, design, and location of homes and schools also have a significant impact the exposure of children to toxicants and other physical hazards. Thus, children's health can be enhanced by the proper design, maintenance, and management of neighborhoods, buildings and others spaces that children inhabit and frequent (4).

Disease prevention and health promotion activities targeting children must also reflect the realities of the physiology and behavior of children. For example, activities focusing on toxicant exposure must recognize the fact that children are more susceptible and vulnerable to environmental toxicants. Compared with adults, children absorb more environmental toxins through the skin because of their greater surface-area-to-body-mass ratio. Their larger air intake relative to body size and closer proximity to the ground increase their vulnerability to air pollutants such as car exhaust. They also have a greater tendency to touch and ingest potentially toxic objects they find in their environment. Because of their faster growth and a higher metabolic rate, they are more likely to incorporate such toxins into their maturing organ systems and suffer health consequences some of which may only become evident much later in life. For all these reasons, children require special attention and cannot be treated as if they are simply small adults.

Effective child health and wellness promotion also requires a realistic assessment of psychosocial, political, cultural, and economic factors that impact children's access to healthcare and other resources. Children who grow up in circumstances of social and economic disadvantage are especially likely to experience health problems because of where they live, learn, and play. They are also less likely to have their healthcare and educational needs met. As a result, they are prone to become trapped in a multi-generational cycle of environmental health disparities that interferes with their ability to reach their full potential (5-7).

The cycle of environmental health disparities

Nurturing and resource-rich environments promote optimal health and development for children. Unfortunately, children living in circumstances of social and economic disadvantage often lack adequate care and nurturing. Instead, they tend to live in unsafe and unsupportive environments that place them at increased risk for environmental factors that result in poor health, developmental delays, impaired learning, and other difficulties that reduce their overall quality of life and limit opportunities for success in adulthood (8-11).

In 2005 the ISDD launched a project, known as *Break the Cycle of Disadvantage and Disability*, which introduced a holistic model that explained how economic, social, psychological, and physical factors compounded the challenges faced by individuals with limited resources. It also emphasized the importance of 'breaking' the cycle of disadvantage and disability that perpetuates such difficulties, both over the course of a lifetime and across generations (12). At the same time, the experiences of the SE PEHSU working on the Vision 2020 project in Anniston, Alabama underscored the importance of adopting a comprehensive approach to children's environmental health that includes recognition of how environmental health disparities are shaped by social and economic factors and perpetuated over time (11). The *Break the Cycle* model was then readily adapted to the bigger picture of children's environmental health disparities and reformulated as the *Cycle of Environmental Health Disparities* (see figure 1).

As the model illustrates, children living in families headed by adults with limited education and health literacy, limited income and employment opportunities, and a relative lack of power within their community, are at increased risk for experiencing adverse environmental health outcomes. Their experience of social and economic disadvantage often leaves them with few housing options in rural, urban, or suburban communities. They are more likely to live in poorly-maintained older homes and be exposed to indoor toxicants, such as lead paint and mold, as well as other outdoor environmental hazards. They are also more likely to live in communities with limited access to transportation and limited availability of grocery stores, green space, and recreational opportunities in their immediate environment. Moreover, they are more likely to be exposed to physical risk and violence in their neighborhoods.

The increased risk of toxicant exposure, poor nutrition, limited physical activity, abuse, neglect, and stress that are often shaped or exacerbated by environmental conditions, in turn increase children's risk of physical health problems, such as asthma, obesity, hypertension and neurodevelopmental disorders. They may also interact in a complex manner that contributes to and compounds learning disabilities, ADHD, depression, trauma reactions, substance abuse, and behavioral disorders. A lack of access to comprehensive healthcare and appropriate educational services, as well as discrimination, social stigma, and lack of social capital further increase the likelihood that these children will continue to experience adverse health outcomes, even into adulthood.

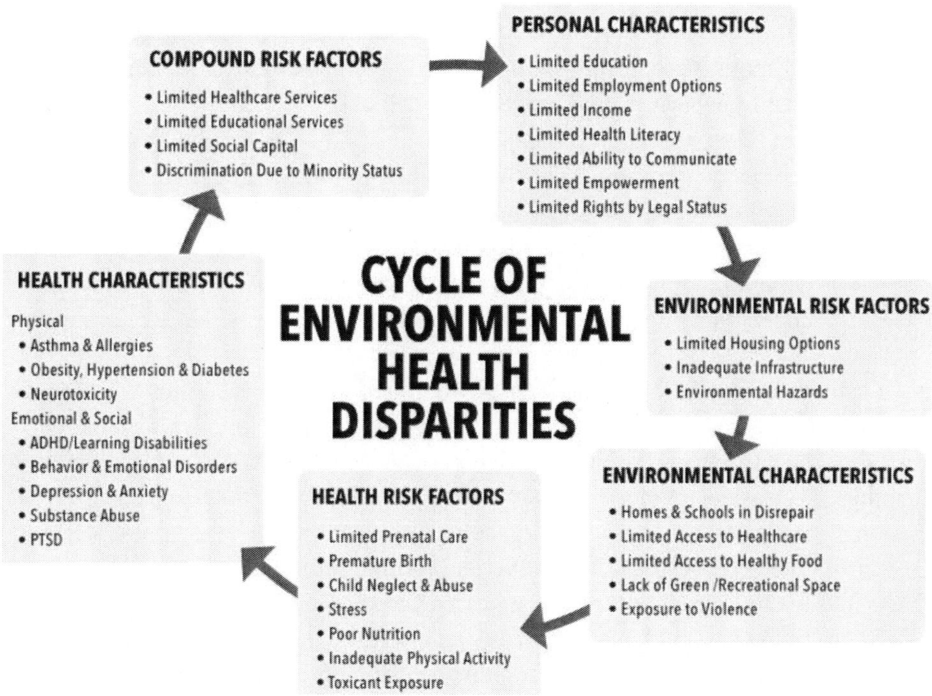

Figure 1. Cycle of environmental health disparities.

Break the cycle

The *Break the Cycle* program is a collaborative, interdisciplinary pediatric environmental health research and training program that encourages university students supervised by academic mentors to conduct research related to the reduction or prevention of environment-related illness and disability in children living in circumstances of social and economic disadvantage. *Break the Cycle* is designed to:

- Inspire students from a variety of academic disciplines to explore the relationship between the environment and children's health, as well as develop strategies for addressing identified children's environmental health disparities
- Promote collaboration among academic leaders at major universities to facilitate creative examination of the issues relating to the impact of environmental factors on children's health and quality of life particularly in relation to social and economic determinants of health
- Promote the incorporation of topics elted to children's environmental health disparities into university curricula
- Develop the leadership capacity of professionals within academic and healthcare communities who can, in turn, promote interest in children's environmental health and raise challenges of environmental health disparities and environmental justice
- Cultivate future leaders and advocates in our societies by inspiring and supporting students to undertake the challenegs that we face today

Students and faculty mentors are recruited from universities throughout North America and countries in South America where relationships have been established. Proposed projects are reviewed for relevance, design, and fit within the project cycle by a committee comprised of the *Break the Cycle* team, representative university faculty, environmental and public health experts. The selection is competitive and a finite number of student projects is selected for that year. During the research phase, monthly conference calls are held to track the students' progress and ensure that students and mentors receive sufficient guidance and support on their projects as well as how their projects relate specifically to the concept of the cycle.

At the end of each project period, the student researchers are required to present their findings at a symposium sponsored by the SE PEHSU, ISDD, and other strategic partners. This symposium not only provides budding scientists, academics and advocates with an opportunity to hone their research presentation skills, but also allows for dissemination of their findings to other scientists and academics as well as healthcare providers, fellow students, advocates and the general public. Each symposium features a nationally-recognized keynote speaker selected for the ability to inspire participants and promote relevant discussion. At the conclusion of each project cycle, the projects are written as scientific papers and published as special issues in peer-reviewed international journals (12-17).

A selection of papers from the past student projects that focused on the theme of home, school and community environments has been assembled here to illustrate the work of the students, the ideas that are embodied in the work and the way in which we can make a positive difference for society's most vulnerable children where they live, play and learn.

Conclusion

Children living in circumstances of social and economic disadvantage are at increased risk for experiencing health problems caused or exacerbated by environmental factors. They are not only vulnerable to adverse chemical and physical factors in their environment, but often become trapped in the cycle of environmental health disparities due to low parental health literacy, limited social capital, and a lack of access to comprehensive healthcare and appropriate educational services. They will only be able to escape that cycle when children's environmental health becomes a priority and when available resources, energy and creativity are amassed and allocated to address elevant environmental health concerns. Such investment may seem remote, given the current economic realities. However, not investing in children's environmental health will prove costlier in the long run.

ISDD and the SE PEHSU have demonstrated that *Break the Cycle* is a sustainable, cost-effective catalyst for increasing academic interest and advocacy in issues relating to promotion of children's environmental health and in reducing health disparities for children. The emphasis of the *Break the Cycle* program is on recruiting graduate and undergraduate students to become aware of environmental health disparities and cultivate their intellect, interest and creativity to focus on a single project that will make a difference. This encourages the incorporation of information about children's environmental health and environmental health disparities into college curricula. It also encourages students and faculty to embrace a broader, more holistic understanding of children's environmental health and environmental justice. Additionally, *Break the Cycle* affords academic mentors who are committed to issues of children's health and the environment, environmental health disparities and environmental justice, with an the opportunity to inform, guide, and inspire the next generation of professionals who will face the challenge of finding creative solutions to environmental health dilemmas. By helping shape such future leaders, this program furthers the goal of breaking the cycle of children's environmental health disparities to promote and assure safe and healthy environments for all children to grow and develop to their full potential.

References

[1] Etzel, RA, Balk, SJ, eds. Pediatric environmental health, 3rd ed. Elk Grove Village, IL: Am Acad Pediatrics Comm Environ Health, 2012.

[2] Paulson JA, Gitterman BA. Children's health and the environment: Part I. Pediatr Clin North Am 2007;54(1):1-212.

[3] Paulson JA, Gitterman BA. Children's health and the environment: Part II. Pediatr Clin North Am 2007;54(2):213-424.

[4] Frumkin, H, Frank, L, Jackson R. Urban sprawl and public health: Designing, planning and building for healthy communities. Washington, DC: Island Press, 2004.

[5] Paulson JA, Karr CJ, Seltzer JM, Cherry DC, Sheffield PE, Cifuentes E, Buka I, Amler RW. Development of the pediatric environmental health specialty unit network in North America. Am J Public Health 2009;99:S511–6.

[6] Vision 2020 Organization. Vision 2020: For the children of Anniston. Children's health environmental justice revitalization project. Anniston,AL:Vision 2020 Org, 2002.

[7] Frumkin, H, Geller R, Rubin IL, Nodvin J, eds. Safe and healthy school environments. Oxford: Oxford Univ Press, 2006.

[8] Gee GC, Payne-Sturges DC. Environmental health disparities: a framework integrating psychosocial and environmental concepts. Environ Health Perspect 2004;112:1645.

[9] Hood E. Focus: Dwelling disparities: How poor housing leads to poor health. Environ Health Perspect 2005;113:A310-7.

[10] Lee C. Environmental justice: Building a unified vision of health and environment. Environ Health Perspect 2002;110(suppl 2):141-4.

[11] Rubin IL, Nodvin JT, Geller RJ, Teague WG, Holzclaw BL, Felner EI. Environmental health disparities and social impact of industrial pollution in a community. The model of Anniston, AL. Pediatr Clin North Am 2007;54:375-98.

[12] Rubin IL, Geller RJ, Nodvin J, Ace K, Merrick J. Break the cycle to improve health and developmental outcomes for all children. Int J Child Adolesc Health 2009;2:265-72.

[13] Rubin IL, Geller RJ, Nodvin J, Ace K, Merrick J. Break the cycle of adverse health and developmental disparities for children. Int J Child Adolesc Health 2009;2(3);265-421.

[14] Rubin IL, Ace K, Nodvin J, Geller RJ, Marcus M, Merrick J. Vulnerable children: Break the cycle of environmental health disparities. Int J Child Health Hum Dev 2010;3(4):357-453.

[15] Rubin IL, Nodvin J, Geller RJ, Marcus M, Merrick J. Vulnerable Children: Break the Cycle of Environmental Health Disparities. Rev Environ Health 2011;26(3):135-229.

[16] Rubin IL, Geller RJ, Nodvin J, Marcus M, Howell M, Merrick J. Break the cycle of environmental health disparities in vulnerable children. Int J Disabil Hum Dev 2012;11(4):295-409.

[17] Rubin IL, Geller RJ, Nodvin J, Marcus M, Howett M, Merrick J, eds. Break the cycle: Environmental hazards and children. Int J Disabil Hum Dev 2013;12(4):in press.

In: Environmental Health
Editors: IL Rubin and J Merrick
ISBN: 978-1-62948-155-5

Chapter II

Use of undergraduate curriculum as a vehicle for breaking the cycle of environmental health disparities within disadvantaged communities

Dominique Bibbins, MPH and Joanne Chu*, PhD
Spelman College, Department of Biology, Atlanta and
The Center for Behavioral Neuroscience, Atlanta, Georgia, US

Abstract

Environmental health disparities (EHD) are disease issues that disproportionately affect particular populations due to complex interactions with environmental conditions. Perpetuation of EHD can be viewed as a closed "cycle" in which factors such as poor living conditions and limited health care exacerbate disease states. The purpose of this study was to investigate whether using an undergraduate curriculum that promotes awareness of EHD among students to develop a potential for leadership and empower them to consider a career and community involvement that could help reduce EHD and develop creative ways to

* Correspondence: Joanne Chu, PhD, EcoEthos Solutions, LLC, 1431 Catherine Street, Decatur, GA 30030, United States. E-mail: Joanne@EcoEthosSolutions.com.

"break the cycle". In the fall of 2007, sophomore and senior students in the biology department at Spelman College examined a case study of polychlorinated biphenyl (PCB) contamination on an African-American community in Anniston, Alabama, a landmark case of EHD. Students were interviewed following the conclusion of the course regarding their attitudes and awareness of EHD.We predicted three qualitative outcomes from our study: 1) students would show increased interest in health disparity careers, 2) students would show increased interest in non-medical environmental health disparities careers, 3) students would show increased awareness/interest in community involvement with environmental health disparities issues. Data from our interviews supported some of our predictions. Implications for future use of undergraduate curriculum to promote interest in the cycle of environmental health disparities are discussed.

Introduction

Health disparities are health and disease issues that disproportionately affect one population over another. Often these populations can also be classified by physical factors such as gender, race, disability or genetic predisposition. However, non-physical factors can also be powerful antecedents that influence disease incidence and prevalence. These factors include socioeconomic status (SES), education level, and living environment (e.g. rural vs. urban areas). Environmental health disparities (EHD) refers to those disease states that are found to be more prevalent in populations defined by disadvantaged social and economic factors, and are often negatively compounded by minority status (1). Unhealthy environments such as chemical plants and hazardous waste facilities are often found in minority and poor communities. Race and ethnicity are often more dominant factors than socioeconomic status when predicting the location and distribution of hazardous waste facilities in the United States. Therefore, racial and ethnic identity are often a critical factors in discussions about EHD (2).

Rubin et al (3) postulated a hypothetical model that illustrates the potential interaction of social, economic environmental and disease variables in a "Cycle of Environmental Health Disparities" (CEHD) (Figure 1). The CEHD model is a closed cycle diagram that depicts how health risks, compounded with social, economic and educational inequalities can feedback to influence community health, healthy development, and growth especially in children.

The cyclical nature of the model suggests that if unbroken, the cycle perpetuates among disadvantaged populations in society.

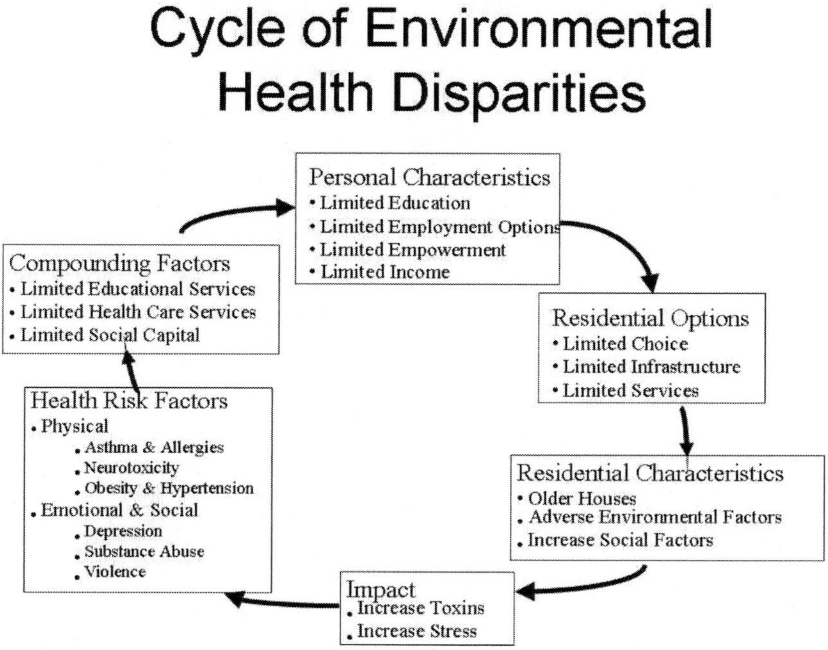

Figure 1. A model for the cycle of environmental health disparities taken from Rubin et al. that describes a hypothetical relationship among multiple environmental variables and health [3].

In this chapter, we present a case study of how the cycle could be "broken" through the use of undergraduate biology curriculum, a unique vehicle for increasing awareness of environmental health disparities among college students and influencing students' career interests in professions that directly or indirectly could interrupt the CEHD. This present study features a departmental seminar course taught in the Biology Department at Spelman College in Fall 2007 entitled, "Sophomore/Senior Seminar – We Live in the World We Make - A Case study of PCB Contamination in Anniston, Alabama". This seminar used a variety of pedagogical approaches including, student-led small and large group discussions, invited seminar speakers, a town hall forum, written reflections and various readings to analyze this landmark case of environmental health disparities and explore its impact on the community and health of a predominantly lower SES, African-American population in Anniston, Alabama.

In this present study we qualitatively investigated the impact of this course on student attitudes and reflections towards environmental health disparities. We hypothesized that this course would impact students in multiple ways, from changing their beliefs and attitudes about environmental health issues, to becoming empowered to help break the CEHD through changes in future professional and personal behaviors. Specifically, we identified three points within the CEHD that female undergraduate biology students of African descent would be especially influential. We hypothesized that this course could "break" the CEHD at multiple points along the cycle (see figure 2):

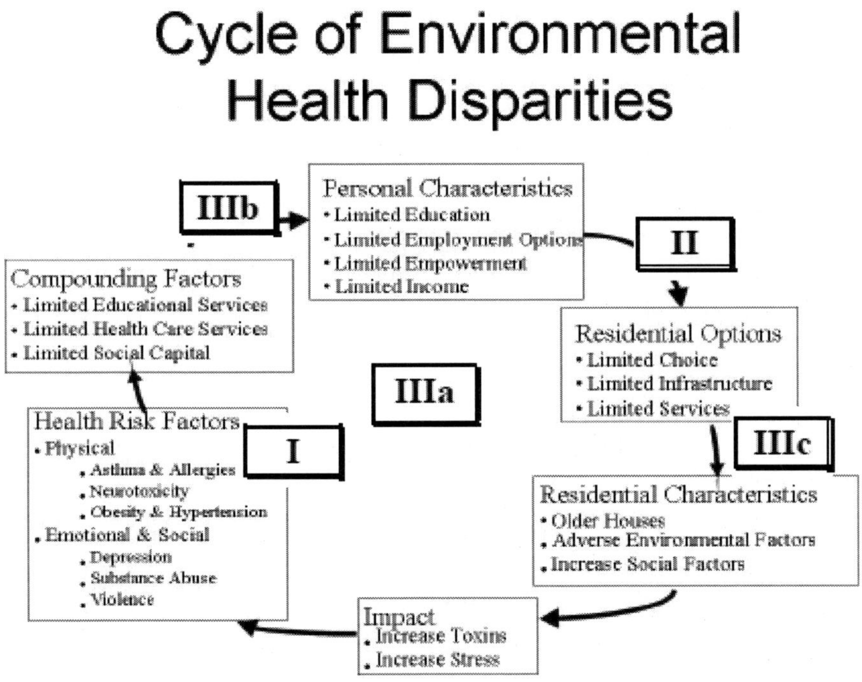

Figure 2. A proposed model in which undergraduate curriculum can help "break" the cycle of environmental health disparities at multiple junctions of the cycle (I, II, IIIa, IIIb, IIIc).

- Influencing Interest in Medical Professions in Health Disparities Areas. Pre-health biology students would show increased interest in medical fields that focus on health disparities health risk factors such as cancer, diabetes, hypertension, etc. (see figure 2, I)

- Influencing Interest in Non-medical Professions in Health Disparity Areas. Other biology students interested in a broader range of careers would show interest in careers in law, public policy and environmental justice. These careers could focus on improving the physical environment of disadvantaged communities (see figure 2, II)
- Influencing Interest in Community. Students, regardless of career interest would show an increased awareness of how changes in community involvement could help break the CEHD. These changes in awareness would come from three different levels of empowerment:
  - The Power of the Individual. Students would better understand how their own behaviors can help break the CEHD through their own involvement in their families, volunteer work, and church groups. Changes in individual behavior have the potential to impact all points of the cycle (see figure 2, IIIA)
  - The Power of Colleges and Universities. Colleges and universities could help break the CEHD by shaping the ways students view the complexities of environmental health disparities through the use of curriculum. Colleges could directly impact the personal characteristics of populations of students through educational interventions (see figure 2, IIIB)
  - The Power of Community. Churches and other community organizations can impact the cycle through increasing awareness of the environmental health disparities within their own communities. Community organizations can impact the cycle by impacting the environmental and social factors of disadvantaged communities (see figure 2, IIIC).

The case: The history and legacy of PCB contamination in Anniston, Alabama

Anniston, Alabama, a small town of approximately 25,000 residents is located in northeast Alabama, 80 miles east of Atlanta, Georgia. Beginning in the 1930's, a chemical plant located in Anniston began producing polychlorinated biphenyls (PCBs), compounds that were widely used for several decades as an industrial coolants and lubricants. In 1977, the production of PCBs was banned in the United States due to increasing evidence of chemical accumulation in the environment and harmful health effects in laboratory animal studies. During the height of PCB production, the plant was dumping several hundred pounds of PCBs into a nearby creek daily. The toxins leached into the ground water and runoff contaminated the soil. Even after the production of PCBs was officially banned, court documents revealed that PCBs were still being released into local environment for many years. In the 1970's through the 1980's there was little efficacy from state and federal agencies attempts to enforce the company to implement environmental remediation action. Finally, in the mid 1990's Anniston residents became aware of the toxins that were polluting their environment and began to ask questions about whether poor health conditions, developmental disabilities, cancers and untimely deaths in their community might be related to the decades long dumping of PCBs. Although the racial composition of the city of Anniston is over 80% Caucasian, the population that was most affected by the PCB contamination was predominantly African-American of lower socioeconomic class.

Anniston, Alabama and its PCBs were chosen for this class as a case study for several reasons compelling to the students in the course. Its litigation, community activism and public health policy legacy were of national stature; however its impact on students was more so as its story was recent and local. Furthermore, nearly all of the invited seminar speakers as part of this course were professionally involved with this case.

The course structure

In the Fall semester 2007, approximately 135 students and the faculty of the biology department at Spelman College were introduced to Anniston, Alabama and its long history with PCB contamination in a departmental seminar course. This course met for one hour per week for the entire semester and was mandatory for all biology majors of sophomore and senior class standing. This course was also a departmental capstone experience for all graduating students. As such, seniors were asked to undertake leadership roles throughout the course.

In the first few weeks of the semester, students were asked to read articles from various popular (non-technical) sources. These included newspaper articles from the Anniston Star (the daily Anniston newspaper), New York Times, Washington Post and an environmental justice website (see table 1 for reading list) (4-8). Each student was assigned one article and was asked to bring that specific perspective to small group discussions of 15 students each. Each discussion group was facilitated by a team of two seniors that had access to all of the readings. The facilitators were charged with helping their group assemble the various narratives of the case study and to initiate class discussion and interest. Although a faculty member was present in the room, all discussion groups were student driven, which allowed students to be more engaged with each other and open to discussion.

Table 1. Readings Assigned to Students as an Introduction to the Anniston PCB Case Study

Article Title	Year	Source
Poisoned By PCBs: Thirty Years Later, Court Documents Reveal Monsanto's Toll on an Alabama Town	?	http://www.chemicalindustryarchives.org/dirtysecrets/anniston/1.asp
Pollution Drives Away Neighborhood and Trust	1997	Bragg, R (1997, March 16), *The New York Times*
Tainted History	2000	(2000, December 10), *The Anniston Star*
Monsanto's PCBs to be here long after the damage Spreads beyond our borders	2001	Op-ed (2001, February 25), *The Anniston Star*
Monsanto Held Liable for PCB Dumping	2002	(2002, February 22), *The Washington Post*

Following this introduction, the remainder of the semester consisted of student team presentations and outside speakers most of whom were involved with the Anniston case. The week before each speaker's arrival, a host team of three seniors presented to the class the speaker's educational background and professional trajectory, focusing on the major stages of their professional development and involvement with the Anniston case. After this presentation, the class generated a set of questions to be posed to the speaker the following week.

Each seminar speaker spoke briefly to the class, from professional prospectives and other perspectives they brought to the case, such as gender, racial or cultural factors. Following these comments, the host team interviewed the speaker using the questions generated by the class in a town hall format.

The class were also allowed time to ask any additional questions. Students' questions ranged widely, from further inquiries about the case, professional trajectories of the speaker, to how the guests' various perspectives helped inform him/her further about case. The guest speakers for this course were:

- Jewell Harper, JD, Deputy Assistant Regional Administrator of the Environmental Protection Agency Region 4, who presented the perspective of the EPA and a legal perspective of the case.
- Leslie Rubin, MD, Co-director of the Southeast Pediatric Environmental Health Specialty Unit, who presented the pediatric environmental health disparities policy perspective of the case.
- Shirley Baker LPN, President and CEO of Mothers and Daughters Protecting Children's Health, who presented the citizen and community activist perspective of the case.
- Stephanie Miles-Richardson DVM PhD, of the Centers for Disease Control and Office of Minority Health and Health Disparities who presented the epidemiological perspective of the case.

In addition to class participation, students were also evaluated on a few short writing assignments. One essay asked students to write about the speaker they found to be most intellectually engaging and reflect on how the speaker's various perspectives helped inform him/her about the Anniston case. Students were also asked to reflect on how that speaker helped changed the student's understanding of the case. This essay also asked the students to reflect on how their own multiple identities help inform their understanding of the case.

Our study

Spelman College is a small, historically black, liberal arts college for women located in Atlanta, GA. The College averages a total student enrollment of 2,100 students. Nationally, Spelman ranks highly in its success in sending underrepresented women to graduate or professional school in the sciences and health professions. The largest major among the sciences is biology, with approximately 350 total declared majors (9). Over the last few decades, the biology department at Spelman has actively sought to modify its biology curriculum to better address current trends and changes within the biological sciences, societal demands for life science professionals, and a changing student population (10,11).

Study design

A set of 15 open-ended questions interview questions was developed by a team of two sophomores and two seniors from the course. These questions were designed to qualitatively examine how this course impacted students in their understanding and awareness of environmental health disparities and to assess whether this awareness could impact attitudes regarding future personal behaviors or professional careers. These questions were based on how the student team felt the course impacted their own lives following this course. These students agreed that a course such as this could "break" the cycle at three major junctions:

- By increasing student awareness in medical careers that focus on health disparities,
- By increasing student awareness of non-medical careers that a biology student could pursue such as law, public policy or environmental justice careers
- By increasing student interest in community involvement through changes in personal behaviors, through educational interventions, and other forms of community involvement (see figure 2, III A – C, respectively)).

Main study questions

The interview questions assessed the following five issues:

- Student knowledge and awareness of environmental health disparities before and after the course.
- Student knowledge of the relationship between the environment and their chosen career paths before and after the course.
- Student assessment of how the course could impact the CEHD.
- Student assessment of how Spelman College or other communities they belong to could impact the CEHD.
- Student assessment of how their own personal behaviors could impact the CEHD.

Interview Process

The team was divided into one pair each of sophomore and senior interviewer pairs. Students from the course were asked to volunteer to be interviewed for this project. Each pair interviewed three students across two sessions (N = 12 total). Students only met within peer groups, sophomores and seniors were interviewed separately. For part of the interview procedure, students were shown a diagram of the CEHD model taken from Rubin et al. and asked several questions regarding how they or their communities could "break" the CEHD (3). All of these interviews were recorded with digital voice recorders so that responses could be transcribed and applied to the CEHD model. The policies and procedures used in this study were approved by the Institutional Review Board for use of human subjects at Spelman College.

Findings

The results of our study are presented in qualitative form as representative quotes extracted from the interviews. Each quote represents a statement made by an individual student in response to one of the interview questions. Repeated or similar quotes are not represented in the tables. Tables show qualitatively different responses only.

Attitude and beliefs about EHD

Our interviews indicated that prior to taking the course, students showed little knowledge or awareness about the concept of EHD (see table 2). Following the course, students showed increased awareness of EHD and of environmental issues in general (tables 3-4).

Changes in career interests and future behavior with regards to EHD: Our interviews showed that prior to this course, students had little awareness of how environmental issues impacted career interests in health or medicine, although some students had awareness of health disparities. Following this course, students showed increased awareness of how environmental issues intersected with their pre-existing interests in health and medicine (table 3). Few students showed an increase in non-health professions related to EHD, although they showed an appreciation for the different professional perspectives presented throughout the semester (table 3).

Table 2. Student Responses - Knowledge of Environmental Health Disparities

Attitudes and Beliefs of Environmental Disparities Prior to Course:

I had very little knowledge of environmental health disparities.

I was not very familiar with the term environmental health disparities.

I knew . . .about environmental health disparities from what I saw on the movie Erin Brockovich.

I never really thought about it, but I would guess that [the environment and health] would have some kind of relationship.

Attitudes and Beliefs of Environmental Disparities Following Course:

Yes, [my awareness was raised] in many places. . . and I learned many things that I never knew about.

Just by raising awareness of the importance of being educated about the environment

I am more conscious about the environment that I live in, and I think that I will carry that into all aspects of my life.

Table 3. Student Responses – Awareness and Interest in Careersin Health Disparities Areas

Influencing Interest in Medical Professions in Health Disparities Areas

I still want to go to medical school . . . but now I feel like I am at an advantage because I know how the environment may be related to certain health disparities.

When I am a doctor, I plan to focus on people who suffer from environmental health disparities

Now I would like to look into careers that deal with environmental health disparities from a public health perspective

I can impact [the CEHD] as an aspiring physician in . . factors such as limited access to healthcare due to limited clinics and health insurance

Understanding health disparities is important as an aspiring physician because some diagnoses may be related to a poor living environment"

Influencing Interest in Non-Medical Professions In Health Disparities Areas

The course helped me a lot because I didn't know what I wanted . . . as a career. . . I've always been a person geared towards the community and community service. After taking this course, I realized that a career in public health that focuses on health disparities would be perfect for me."

I liked the fact that we had different people come in, like a lawyer and a nurse/ community activist. It was good to see how biology related to these different fields

Table 4. Student Responses – Awareness and Interest in Community Involvement

Influencing Interest in Community
A. The Power of the Individual
I would be more conscious about how the environment affects our health. Instead of only looking at the lifestyle and eating habits of my patients as it relates to health disparities, I will also look into the environmental factors that may be adversely affecting their health.
Awareness is one of the most important things, because once you inform other people, they will inform others, and this will eventually break the cycle "
I already have [an increasing interest in EHD]. . . I will continue . . .[to tell] friends and family members about the implications of living in a poor environment.
B. The Power of Colleges and Universities
Spelman has already begun impacting the CEHD. By having classes such as [this] . . . they are educating us on the environment, which allows us to disseminate the information we learned to others.
Spelman has already begun to impact the cycle by building the first ever green dorm at a historically black college. This has opened dialogue and raised awareness about the importance of being environmentally friendly.
Spelman and other institutions like it can begin to impact the cycle by reaching out to the community and raising awareness about environmental health disparities
Institutions like Spelman can implement courses like these into other disciplines and majors. It can be like a bridge between the sciences and non-sciences
Having courses like this in other disciplines could help [break the CEHD]
C. The Power of Community
Community activists, teachers and church leaders can impact the cycle by raising awareness of EHD within their neighborhoods, schools, and churches
When people become more aware, they become more empowered...when people gain more knowledge, nobody can take advantage of them. That's why raising awareness within the community is so important

Students had multiple responses for how their personal and community behaviors could impact the CEHD (table 4A-C). Responses from students reflected their awareness they already belong to communities that could raise awareness of and intervene in the CEHD. Responses were strongest as to how students viewed how their own personal behaviors could impact the CEHD (table 4A) and how institutions, like Spelman College (table 4B) could impact the cycle. Views on how their communities could help disadvantaged communities seemed less strong (table 4C).

Discussion

The original intent of this departmental seminar was not to focus on the issues of EHD per se. The use of the Anniston, Alabama case study was used as a means to illustrate to students the complexities of real world scenarios where biology can inform the multiple perspectives of this case. Our study, however, did examine whether students showed increased awareness of environmental disparity issues and increased interest in how they might be empowered to "break the cycle of environmental health disparities" following this course. Our interviews qualitatively indicated that students did engage with the course topic on issues of EHD.

Our study indicated that this undergraduate seminar course taught using Anniston, Alabama, and its history and legacy of PCB contamination did impact students' awareness and attitudes about EHD. Specifically, we were interested in whether this course could impact students' interest in professional arenas related to EHD. The vast majority of biology students at Spelman College are interested in health professional careers. Our interviews indicated that among students interested in health careers, students showed a greater awareness of how environmental factors disproportionately impact disadvantaged populations and indicated a greater interest in those health areas subject to disparities such as cancer, diabetes and cardiovascular illnesses. We had also hypothesized that students would show an increased interest in non-medical fields, such as law, public policy and environmental justice careers. Our interviews did not support this second hypothesis. However, in examining student written work, a slightly different story emerges. Approximately 100 students out of the course chose to compose an essay in response to the prompt: "Describe the speaker's professional identity that you found to be most intellectually interesting and explain why it was attractive to you". The majority of students chose to reflect upon Jewell Harper, the EPA administrator who presented the legal aspects of the case. The students' next choice was Leslie Rubin of PEHSU, the physician who presented the medical and public health policy perspectives of the case (data not shown). The students' first choice was slightly surprising, as Spelman biology students have not historically expressed interest in law or careers with federal agencies such as the EPA.

Our final hypothesis was that students could make a connection between the CEHD and their own involvement in the community. Students in our interviews seemed to possess an awareness of how their own personal knowledge of EHD could help "break" the CEHD. Even stronger were the responses to questions about the role of the college in its potential role to help impact the cycle. Students were very clear that college curricular or community objectives could have a strong impact on helping "break" the CEHD. Students were less specific of how community involvement through other institutions through church or other volunteer work could impact the cycle, although their statements are consistent with studies that show that African-American churches are more actively involved in serving underprivileged communities than White churches (12) and with students self-reported value for community service in their lives (pers. communication). Overall, students felt strongly that the CEHD could be impacted through more active involvement through personal, educational and community activities.

Undergraduate curriculum is a novel approach for increasing awareness in EHD issues, especially in the liberal arts. In this case, Biology majors in this course were engaged to consider EHD issues within their own career choices and community activities. By using undergraduate curriculum as a vehicle to help "break" the CEHD within disadvantaged communities, courses such as this can help shape the way students view the environment and health, and empower students to intercede in this ongoing cycle.

Education scholars have long advocated that undergraduate science curriculum be less content driven and more focused on skills and critical thinking development. The rapidly changing world of science makes it virtually impossible to teach more than a selection of facts. Emphasis on skills development and critical thinking better prepares today's student to acquire knowledge in tomorrow's world. Modern pedagogy places emphasis on learning that is student-centered and active. Active learning places more focus on the student to be responsible for the learning experience and is less focused on the professor as the main authority in the classroom. Thus, professors are transformed from their traditional roles as the "Sage on the Stage" and become more the "Guide on the Side" (11). Students that take an active role in their own learning have a greater ownership over that knowledge and that knowledge has a greater potential to impact future behaviors and decisions (13).

Several techniques of student-centered learning were used in this course. One main technique was the use of a group work in conjunction with a case-study approach. Students in traditional settings are often inundated with acquiring facts they often do not fully understand or cannot commit to long-term memory. Presentation of facts without context leads to rote, short-term memorization. Case studies in the classroom can be an excellent framework for student-centered learning as the learning takes place within a cohesive narrative (14). In addition to setting the learning experience in context, cases also provide compelling means for students challenge their own understandings of the world and to gain knowledge about the world they live in. This "real world" knowledge can help shape and inform their world views and may help guide future career directions. One way to probe student engagement and understanding of the case study material is to provide ample opportunity for group discussion. Group discussion provides a forum for students to extract the information they need. Through group activities these students were responsible for generating their own understanding of the Anniston case through discussion and questions directed at student presenters, classmates and seminar speakers. We believe that this course framework in conjunction with the course topic were important factors in creating a lasting impact for students to become aware of environmental health disparity issues and possibly change their future behaviors.

We believe that undergraduate curriculum could be an effective means to impact the CEHD and to encourage students to think proactively about their future careers and personal involvement in larger community issues. This preliminary study should encourage professionals in EHD fields to discuss with college faculty and administrators the impact curriculum has on student engagement with health disparities and student interest in careers that might ameliorate these social injustices. Professionals in these fields should collaborate with academic faculty to think creatively of ways to engage students in both science and non-science disciplines to discuss EHD in the classroom. In addition to biology, these disciplines could easily include such areas as chemistry, mathematics, political science, women's studies and history. Ultimately, we hope that the information from this preliminary study will help generate additional ideas using undergraduate student populations and undergraduate curriculum to contribute towards, "breaking the cycle of environmental health disparities".

Acknowledgements

Thanks to Dr. Cynthia Bauerle for help developing and executing the course; Dr. Latanya Hammonds-Odie for help developing the course framework; the student team – Brittany Benson, Amirah Patterson and DeAnna McGarity for help developing the study; Rob Poh for the audio portion of this study. This work was supported by Howard Hughes Medical Institute Undergraduate Education grant to the Biology Department at Spelman College for curricular development, the Center for Behavioral Neuroscience, Atlanta, GA and the Southeast Pediatric Environmental Health Specialty Unit (PEHSU). A special thanks to Dr. Leslie Rubin co-director of PEHSU for his encouragement and unflagging enthusiasm for the seminar course, Spelman students and this study.

References

[1] Gee GC, Payne-Sturges DC. Environmental health disparities: a framework integrating psychosocial and environmental concepts. Environ Health Perspect 2004;112(17):1645-53.

[2] The quest for environmental justice: Human rights and the politics of pollution. San Francisco, CA: Sierra Club Books, 2005.

[3] Rubin IL, Nodvin JT, Geller RJ, et al. Environmental health disparities: environmental and social impact of industrial pollution in a community - the model of Anniston, AL. Pediatr Clin North Am 2007;54(2):375-98.

[4] Bragg R. Pollution drives away neighborhood and trust. New York Times 1997 Mar 16.

[5] Grunwald M. Mansanto held liable for PCB dumping. Washington Post 2002 Febr 22.

[6] Myers J. Monsanto's PCBs to be here long after the damage spreads beyond our borders. Anniston Star 2001 Mar 29.

[7] Raeke R. Tainted history. Anniston Star 2000 Dec 10.

[8] Poisoned by PCBs: Thirty years later, Court documents reveal Monsanto's toll on an Alabama town. [cited; Available from: http://www.chemicalindustryarchives.org/dirtysecrets/anniston/1.asp

[9] Office of Institutional Research AaP. Atlanta, GA: Spelman Coll Fact Book, Spelman Coll, 2005-2006.

[10] Transforming America's scientific and technological infrastructrure: Recommendations for urgent action. Washington, DC: Project Kaleidoscope, 2006.

[11] King A. Changing college classroom new teaching and learning strategies for an increasingly complex world. San Fransicso: Jossey-Bass, 1994.

[12] Chaves M, Higgins LM. Comparing the community involvement of Black and White congregations. J Sci Stud Religion 1992;31(4):425-40.

[13] How students learn: History, mathematics, and science in the classroom. Washington, DC: Natl Acad Press, 2005.

[14] Herreid CF. Start with a story: The case study method of teaching college science. Arlington, VA: Natl Sci Teachers Assoc Press, 2006.

Section 1: The Home

In: Environmental Health
Editors: IL Rubin and J Merrick
ISBN: 978-1-62948-155-5

Chapter III

Planning for healthy homes

Lindsay Ann Herendeen, MPH and Amy MacDonald*, MS
Gillings School of Global Public Health,
Health Behavior and Health Education
University of North Carolina Chapel Hill, Durham and
NC Childhood Lead Poisoning Prevention Program Coordinator,
Institute for the Environment, Environmental Resource Program,
University of North Carolina Chapel Hill, Chapel Hill,
North Carolina, US

Abstract

Substandard housing conditions have been linked to widespread childhood environmental health ailments, including two of the leading causes of childhood morbidity: lead poisoning and asthma. In 2009, the US Surgeon General called for action around healthy homes. Improving home health environments can alleviate the cycle of childhood morbidity and mortality. The North Carolina Department of Environment and Natural Resources Children's Environmental Health Branch is working to

* Correspondence: Amy MacDonald, MS, Environmental Health Educator, NC Childhood Lead Poisoning Prevention Program Coordinator, University of North Carolina Chapel Hill, Institute for the Environment, Environmental Resource Program, 337 W Rosemary St, Chapel Hill, NC 27599, United States. E-mail: amyjmac@email.unc.edu.

build capacity at the state-level to expand the childhood lead poisoning prevention program to respond to additional in-home environmental health issues. To achieve this objective, North Carolina must consider recommendations for assessment, management, and evaluation. This paper will situate healthy homes on the national public health agenda; discuss ways that healthy homes programs address children's environmental health disparities; introduce the NC Healthy Homes Initiative; explore current healthy housing efforts in North Carolina through an examination of the Guilford County Healthy Homes Initiative; and provide recommendations for the NC Healthy Homes Initiative to address children's environmental health disparities.

Introduction

Substandard housing conditions have been linked to widespread childhood environmental health ailments, including two of the leading causes of childhood morbidity: lead poisoning and asthma (1,2). In 2009, approximately 1.2% of US children had blood lead levels elevated above 10μg/dL and 9.4% of children suffered from clinically-diagnosed asthma (3,4). Children are at a disproportionate risk for environmental health disparities due to behavioral patterns and developmental status (1,2). Normal hand-to-mouth behaviors and consumption of air and water disproportionate to body size place children at greater risk of exposure, increasing opportunities for long-term, negative developmental outcomes (1). However, both lead poisoning and asthmatic episodes can be prevented through environmental change (2). Children spend 90% of time indoors, making indoor environmental quality a key concern for health (1).

In 1999, the US Department of Housing and Urban Development (HUD) launched the Healthy Homes Initiative to holistically address indoor environmental quality. Nationally, the healthy homes movement has grown out of the effort to eliminate childhood lead poisoning (5). From 1996-2006, childhood lead poisoning rates fell from 7.6% to 1.2%, indicating capacity to address other home health concerns, such as indoor air quality, pesticide use, and home safety (3). Healthy homes programs have shown that a comprehensive, coordinated approach to environmental heath ailments may generate a greater public health impact by influencing many problems at once, rather than focusing on a single concern (5,6).

The definition of a "healthy home" has evolved over the past decade. Initially, HUD defined a healthy home as one that is dry, clean, pest-free, contaminant-free, ventilated, well-maintained, and safe (7). In action, this initial definition focused on ways to improve the structure of the home in order to improve the health of occupants. Over time, this definition grew to encompass ways that the home is sited, designed, built, maintained, and renovated (6). Focus expanded to target the behaviors of individuals living within the home as well physical, social, and mental health of residents.

In 2009, the US Surgeon General issued a Call to Action around healthy homes. This resulted in a shift in national focus to housing-related health issues and caused HUD to join forces with the US Centers for Disease Control and Prevention (CDC) and US Environmental Protection Agency (EPA) to promote a unified agenda (6). The Call to Action expanded the definition of a healthy home further to include the surrounding neighborhood and community, providing a more holistic explanation of the interaction between the individual, community, and built environment (6). These evolving definitions are similar in the emphasis that substandard housing is linked to myriad health outcomes, including: lead poisoning, asthma and allergies, unintentional injury, pesticide exposure, and others (7,8,9). These definitions suggest that housing is an important determinant of health.

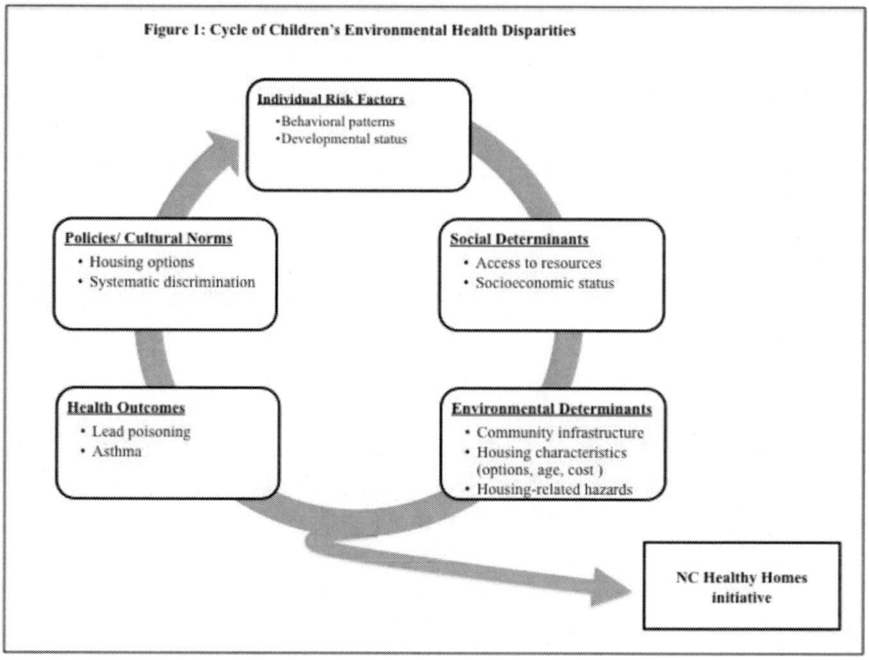

Figure 1.

Social determinants of health associated with housing such as housing options, cost, socioeconomic status, and systematic discrimination, as well as environmental determinants such as structural integrity, hazards, and community infrastructure, have a significant impact on childhood health outcomes (6). Figure 1 illustrates the links between childhood risk factors, social and environmental determinants of health, health outcomes, and political and cultural norms that perpetuate childhood health disparities.

According to a 2009 study conducted by the National Center for Healthy Housing, 67% of families live in a home with at least one major health risk (10). Children are more likely to live in a home with a health risk and in substandard housing if they live in older housing stock, in low socioeconomic status, and in rural communities. Families that live in homes built before 1978, when lead paint was banned, are at a higher risk of exposure to lead (3). Of families that live in a home built before 1980, 73% live below the poverty line (11). In addition, of families that live in a home with water leakage from the exterior, 9.6% live below the poverty line (11). Dampness and mold resulting from water leakage are associated with a 30-50% increase in asthma-related health outcomes (12). Low-income and minority children are more likely to miss school days or become hospitalized for asthma-related health outcomes than other children (13).

Approximately 14 million children live in rural America (14). Rural families are more likely to live in poverty, more likely to be uninsured, and more likely to have poor access to health facilities as compared to families living in urban areas (14). In addition, homes in rural areas are more likely to be built before 1978 and more likely to have degraded structural integrity, safety hazards, and poor overall maintenance (6). Disparities resulting from differential access to safe, affordable housing are further compounded by disparities related to lead poisoning, asthma, and other environmental health concerns.

The Social Cognitive Theory (SCT) states that there is a reciprocal relationship between the individual, behavior, and environment (15). Changes in the environment or in the availability and provision of resources can positively impact behavior change by reducing or eliminating barriers (15). SCT suggests that improving environmental support and eliminating environmental barriers can increase self-efficacy to improve home environmental conditions, helping to facilitate and maintain long-term behavior change (2,15). Individual level changes, in turn, can have long-term impacts on home environments by improving indoor environmental quality, thereby improving childhood health outcomes (2). As most environmental problems supersede the individual level, environmental health interventions require comprehensive, multi-level strategies for health behavior change (16). Healthy homes programs have the ability to break the cycle of childhood environmental health disparities (Figure 1) by altering the link between social and environmental determinants and health outcomes by improving home living conditions. Healthy homes programs seek to address both environmental and behavioral factors by assessing structural home issues as well as individual health behaviors.

This chapter will introduce the NC Healthy Homes Initiative; explore current healthy housing efforts in North Carolina through an examination of the Guilford County Healthy Homes Initiative; and provide recommendations for the NC Healthy Homes Initiative to address children's environmental health disparities.

Literature review

An informal literature review related to healthy homes behaviors, assessment tools, and evaluation was completed. This review evaluated sources from MEDLINE, CINAHL, ISI Web of Science, National Center for Healthy Housing, HUD, and other sources. Articles were identified based on search terms including: healthy homes, housing, lead poisoning, asthma, social change, etc. In addition, a comprehensive analysis of seventeen healthy homes assessment tools used throughout the US was completed (not presented here) to evaluate common concerns addressed by healthy homes programs and the relationship between home environments and individual behaviors (17). A complete review of national healthy homes objectives from HUD, CDC, and EPA provided insight into the role of healthy homes on the national public health agenda (6-9,18).

An exploration of the NC Healthy Homes Initiative was completed through three key information interviews with program staff at NC Department of Environment and Natural Resources Children's Environmental Health Branch (NC DENR) as well as at the Guilford County Department of Public Health and the Greensboro Housing Coalition (19,20). In addition, a review of the NC Strategic Plan to Eliminate Childhood Lead Poisoning, and the mission, goals, and objectives of the NC Childhood Lead Poisoning Prevention Program was completed (21). Outreach work (not presented here) was reviewed by the NC Healthy Homes Outreach Task Force, consisting of statewide program partners and environmental health specialists.

Discussion

In North Carolina, approximately 0.4% of children have elevated blood lead levels and 13% of children suffer from asthma (3,4). Following national trends, North Carolina saw declines in childhood lead poisoning cases from 660 cases in 1997 to 280 in 2007, while increasing screening by 50% (21). In response to this decline in blood lead levels and evidence of the need to address other housing-related health concerns (eg. asthma), North Carolina created the Healthy Homes Outreach Task Force in 2006. In 2008, the Task Force added a healthy homes goal to the NC Strategic Plan to Address Childhood Lead Poisoning to begin the transition from lead to healthy homes (21). This goal aimed "to strengthen initiatives throughout North Carolina that improve the quality of housing, so that children's overall environmental health is protected" (21).

Simultaneously, Guilford County, North Carolina began efforts to create a Healthy Homes program in 2008. As an early-adaptor of the healthy homes model, Guilford County was the first county to conduct healthy homes assessments in North Carolina. Guilford County (population 463,534) is located in the Piedmont Triad region of North Carolina, approximately 30 miles east of Winston-Salem and 100 miles northeast of Charlotte (22,23). Greensboro and High Point are the two largest urban centers; Greensboro is the county seat (23). In 2008, approximately 10.4% of families lived below the poverty threshold, with a countywide median income of $47,308 (22). Approximately 60% of the population is White, 31% of the population is Black, and 6% of the population is of Hispanic or Latino origin (22). Guilford County is comparable to other rural counties in North Carolina (14).

Following attendance at a healthy homes training offered by the National Center for Healthy Housing in 2008, the Guilford County Department of Public Health and Greensboro Housing Coalition, a local non-profit advocacy organization dedicated to safe and healthy housing in the greater Greensboro area, initiated conversations to develop a model for a healthy homes program in Guilford County. The conception of the Guilford County Healthy Homes Initiative was motivated by a desire to pursue an evidence-based, comprehensive program to address children's environmental health concerns; to provide healthy homes training in North Carolina; and to follow shifting trends in federal grant funding and support. Between January 2009 and June 2010, the Guilford County Healthy Homes Initiative completed 270 home visits. Guilford County has developed and implemented the Healthy Homes Initiative while the state has planned the transition to healthy homes.

Throughout this process, NC DENR viewed Guilford County as a leader in the state. The Initiative serves as a model in training, partnerships, visibility, assessment, data management, and evaluation for other agencies in North Carolina at the county and state-level transitioning from lead to healthy homes.

Training

The first objective of the NC Healthy Homes Initiative was to "train health and housing staff and develop an infrastructure to conduct healthy homes assessments" (21). Environmental health specialists from Guilford County and NC DENR Children's Environmental Health Branch attended a two-day, healthy homes training provided by the National Center for Healthy Housing. All environmental health specialists for the North Carolina Childhood Lead Poisoning Prevention Program received healthy homes training over the course of two years. In Guilford County, two staff members at the Department of Public Health are certified to conduct home assessments and three staff members at the Greensboro Housing Coalition serve as dedicated healthy homes staff. Additionally, the Healthy Homes Outreach Task Force hosted speakers on healthy homes topics at quarterly meetings, helping environmental health specialists to gain further knowledge of housing codes, integrated pest management, and in-home hazards.

Well-trained environmental health specialists are a key component of Guilford County's success. As a result of the holistic approach to healthy homes programming, environmental health specialists must obtain an expert and working knowledge of structural, behavioral, and health concerns linked to housing. This requires high levels of training and a commitment from specialists already proficient in childhood lead poisoning prevention to expand into healthy homes topics, such as: indoor air quality, integrated pest management, and home safety. The training of environmental health specialists at the state level ensures that home visit staff will be adequately prepared to evaluate and provide solutions for a wide array of housing concerns.

Partnerships

The Guilford County Healthy Homes Initiative has a unique, strong network of professionals that visit homes and serve populations living in substandard housing stock. These professionals include environmental health specialists, nurses, and maternal-child care coordinators. The Healthy Homes Initiative is primarily a joint effort between the Department of Public Health and the Greensboro Housing Coalition. In addition, the effort also includes partners from non-profit agencies, for-profit agencies, local collaboratives, and universities, including: Cooperative Extension, United Way, YWCA, University of North Carolina Greensboro, Greensboro Healthy Birth Outcomes Task Force, and local organizations (19,20). The coalition works in concert with the Greensboro and High Point Housing Departments, governmental agencies responsible for code enforcement and structural integrity of homes in the two largest urban areas of Guilford County (19). These partnerships are vital to ensure a comprehensive, interdisciplinary, multi-faceted approach to safe and healthy housing.

Like the healthy homes transition nationally, this dense network of partnerships grew out of a lead poisoning prevention coalition consisting of representatives from myriad organizations in Guilford County concerned with children's health outcomes. This coalition facilitated Guilford County's transition to healthy homes by adopting goals related to healthy home environments. The long-standing practice of coalition-building has helped Guilford County in efforts to address housing-related health concerns (20).

As an early-adopter of healthy homes approach, Guilford County is now sharing materials and trainings with other counties across North Carolina to increase state capacity, thereby fulfilling the second objective of the Healthy Homes Outreach Task Force to "recruit partners and allies to enhance healthy homes training and services" (21).

Visibility

The third objective of the Healthy Homes Outreach Task Force, to develop educational outreach materials aimed at educating the general public about the connection between health and housing, was accomplished through the creation of a website, outreach visits, and training opportunities (21). The NC Healthy Homes Initiative website was designed and created by Guilford County, Durham County, and Forsyth County Departments of Public Health, and University of North Carolina Chapel Hill Environmental Resource Program. The website contains a video testimonial detailing a "success story" addressing housing-related health concerns for a child with asthma. The purpose of this story was to model a family seeking healthy homes advice in order to demystify what these visits include and to encourage other families to seek home visits (19). The website also displays the NC Healthy Homes Initiative logo to brand the project.

A key purpose of the NC Healthy Homes Initiative website was to make information and materials accessible to state and local organizations and to the greater North Carolina community. NC DENR hoped to create space for a dialogue between environmental health specialists conducting home visits through resources and tools available on the website (19). In addition to online technology, NC DENR and the Guilford County Healthy Homes Initiative have provided numerous outreach efforts through community programs and trainings for school nurses, childcare providers, and social workers (19). These efforts have helped to increase awareness of the project throughout Guilford County. At the state-level, University of North Carolina Chapel Hill Environmental Resource Program has also developed hands-on healthy homes trainings for environmental health specialists and nurses across North Carolina.

In addition to visibility generated through the website, outreach visits, and training opportunities, NC Healthy Homes Initiative is also in the process of developing and field-testing educational materials for use during home visits targeted to both children and adults. Educational materials emphasize behaviors such as reducing chemicals and eliminating mold and mildew through cost-saving, accessible practices. Initial pilot tests of these materials have identified a need to also assess what materials are useful for home visit staff to explain behavioral and environmental changes, providing an additional opportunity for healthy homes training.

Assessment and data management

NC Healthy Homes Initiative is working to build capacity to offer healthy homes assessments by developing a comprehensive assessment tool for use statewide to accomplish the fourth objective of the Healthy Homes Outreach Task Force: to develop and implement a healthy homes assessment tool and data management system (21). This project completed a review of seventeen publically-available home assessment tools used throughout the US and developed four forms (not presented here), keeping in mind special considerations of the NC DENR environmental health specialists as many are experts in childhood lead poisoning and experience large caseloads and time restrictions. Feedback was received from the NC Healthy Homes Outreach

Task Force at a quarterly meeting to determine which form is most appropriate to capture actual home environmental conditions in North Carolina. Environmental health specialists were encouraged to test the forms in the field and to provide further feedback. NC DENR home visit staff recommended the creation of a checklist with talking points to follow as they progress through a home visit. Environmental health specialists felt that this would help them to feel more confident in recommending behavior changes to improve the home environment, again strengthening training efforts. Specialists also requested a specific section to record applicable childhood health concerns, with the intention that these concerns could be matched to housing conditions. During the transition from lead to healthy homes, this form will be kept brief to add to the end of a lead assessment. The purpose of the form is to allow home visit staff to capture "big picture" concerns so that families can be referred back to NC DENR for tailored home visits.

These assessments tools were designed, in part, based on the form used extensively in Guilford County. The Department of Public Health and Greensboro Housing Coalition use the same forms and data management tools for consistency and to ensure a larger pool of data to obtain resources. Guilford County Healthy Homes Initiative plans to adopt the state forms as soon as possible to increase reporting and tracking consistency statewide (19,20).

With limited resources, Guilford County attempts to collect preliminary data before going into the home to complete an assessment. The Department of Public Health prioritizes home visits for children diagnosed with asthma or showing significant health concerns. A majority of home visits are completed to address mold and mildew concerns linked to childhood asthma. In addition, prescreening also determines whether a family has talked to the Housing Department to ensure that all code violations are addressed before completing a home visit.

Together, the environmental home assessment; the checklist targeting behaviors; and the educational materials start to create a comprehensive program to evaluate environmental and behavioral factors leading to children's health disparities in North Carolina.

Evaluation

As North Carolina prepares to transition from lead to healthy homes, current efforts will position NC DENR to apply for dedicated healthy homes funding from federal agencies to support a comprehensive, statewide healthy homes program. The fifth and last objective of the Healthy Homes Outreach Task Force is to track healthy homes activity and funding opportunities by HUD, CDC, EPA, other state agencies, and non-profit organizations (21). Key to tracking and pursuing funding opportunities is the need to develop a strong evaluation plan. While healthy homes programs have the power to reduce children's environmental health disparities, this can only be assessed if built into the program evaluation. Currently, Guilford County does not complete formal tracking or evaluation, but serves as a proactive, educational program (19). Unlike childhood lead poisoning where it is possible to monitor elevated blood lead levels, it is difficult to measure or evaluate parental empowerment as an outcome of a healthy homes intervention. Without standard enforcement protocols for healthy homes, it is difficult to set evaluation criteria. All current enforcement capacity in Guilford County lies with the Housing Department, who has the ability to cite code violations or substandard housing conditions.

Recommendations

Moving forward with these objectives, NC Healthy Homes Initiative plans to launch a pilot test in 2010 for further field evaluation and assessment of NC DENR home visit staff needs. NC Healthy Homes Initiative has a unique opportunity to pilot and design a strong intervention before implementation. This pilot will be further strengthened by leadership from the Guilford County Healthy Homes Initiative as a model for the rest of the state. In seeking recommendations from Guilford County, NC DENR will be able to transfer strong practices and lessons-learned from Guilford County to other counties, encouraging co-learning and spreading key practices (24).

Guilford County Department of Public Health has a strong partnership with the Greensboro Housing Coalition and housing code inspectors. These partnerships are needed at the state-level and NC DENR must collaborate with all environmental health specialists conducting home visits for children statewide, especially specialists making multiple visits or creating a trusted relationship with families. NC DENR must also look for key statewide partners working for safe and healthy housing (eg. NC Asthma Alliance). State partners should work to develop legislation that will ensure that healthy homes principles are sustainable and enforceable (19,21). Without enforceable standards, it will be difficult to receive support from counties and to construct an effective evaluation and tracking system.

Evaluation components should come from federal leadership, though debates continue about how to rigorously evaluate healthy homes interventions (2,25). At the state-level, NC DENR plans to conduct a follow-up via phone with all families to determine knowledge and behavior changes as well as client satisfaction. NC DENR will continue to compile and publish surveillance data for childhood blood lead testing results and will begin to collect environmental data from lead investigations statewide in 2010. NC DENR will also survey environmental health specialists to determine additional training needs as well as use environmental data collected from pilot assessments to further target home assessments and education. Process measures should also include parental empowerment, behavioral change, and structural change.

Outcome measures must begin to match key childhood health outcomes to home environmental conditions. The CDC and National Center for Healthy Housing have identified key housing interventions found to have the greatest impact on childhood health status (26). Of identified interventions, home environmental assessments and educational outreach form the basis of recommendations (26). NC DENR has reviewed these indicators and selected those that meet the goals and fall within cost parameters of the program. NC DENR will focus on: provision of mattress and pillow covers; use of HEPA vacuums and air filters; reduction of environmental tobacco smoke; management of cockroaches and rodents through practice of Integrated Pest Management; use of safe household cleaning methods to protect indoor air quality; control of lead hazards; installment of working smoke and carbon-monoxide detectors; and use of pre-set, safe temperature hot water heaters. Indicators such as blood lead levels, asthma hospitalization, and school absenteeism may be useful to estimate changes in children's health status resulting from improved housing conditions (2,19,20). The Guilford County Healthy Homes Initiative is seeking partnerships with the medical community and local universities to increase their ability to collect relevant health outcome data and to match this data to census tract-level housing information (19,20).

CDC has recognized the importance of tracking children's environmental health data to evaluate the progress of healthy homes and to address health disparities. The CDC now mandates that all healthy homes project grants include a discussion of ways that environmental justice principles will be incorporated into healthy homes projects (8). This federal leadership will make the link between housing and health disparities more explicit. Questions remain about how to effectively evaluate whether healthy homes programs improve health outcomes to break the cycle of children's environmental health disparities. The NC Healthy Homes Initiative should begin to consider ways to target staff, resources, and funding to target areas of greatest need and to address disparities. As learned from Guilford County, this may be most effective by creating strategic partnerships with agencies that can fill gaps in providing healthy homes outreach, such as faith-based organizations, rural nursing centers, and other non-profit organizations (19).

The NC Healthy Homes Initiative must work to foster statewide partnerships, dialogue, and coalition-building to carefully target services to areas of greatest need; develop assessment tools that capture behavioral, environmental, and structural housing concerns to measure effectiveness; and link children's health outcomes with home environmental conditions. Drawing on the expertise of state environmental health specialists and providing additional training in key areas, home visit staff can offer feasible solutions to improve home environmental conditions for the most vulnerable populations. In this way, the NC Healthy Homes Initiative may have the ability to create long-term, sustained change, effectively breaking the cycle of children's environmental health disparities in North Carolina.

Acknowledgements

This project was completed with support from the Break the Cycle of Environmental Health Disparities research program of Emory University and the Southeast Pediatric Environmental Health Specialty Unit. Special thanks to the Guilford County Department of Public Health, Greensboro Housing Coalition, and NC Healthy Homes Outreach Task Force; Dr. Arjumand Siddiqi, University of North Carolina Chapel Hill, Gillings School of Global Public Health; and the staff of the Break the Cycle program. This chapter is a revised and adapted version of an earlier publication: Herendeen LA, MacDonald A. Planning for the North Carolina healthy homes initiative. Rev Environ Health. 2011;26(3):149-54.

References

[1] Breysse P, Farr N, Galke W, Lanphear B, Morley R, Bergofsky L. The relationship between housing and health: Children at risk. Environ Health Perspect 2004;112(15):1583-8.

[2] Krieger J, Takaro T, Allen C, Song L, Weaver M, Chai S, Dickey P. The Seattle-King County Healthy Homes Project: Implementation of a comprehensive approach to improving indoor environmental quality for low-income children with asthma. Environ Health Perspect 2002;110(S2):311-22.

[3] Centers for Disease Control and Prevention. CDC's national surveillance data (1997-2007): Lead. CDC 2008. Accessed 2013 May 20. URL: http://www.cdc.gov /nceh/lead/data/national.htm.

[4] Centers for Disease Control and Prevention. Summary health statistics for US children: National Health Interview Survey, 2008. CDC Vital and Health Statistics 2008;10(224):9.

[5] US Department of Housing and Urban Development. HUD history. HUD 2009. Accessed 2013 May 20. URL: http://portal.hud.gov/portal/ page/portal/HUD/ about/hud_history.

[6] US Department of Health and Human Services. The Surgeon General's call to action to promote healthy homes. Department of Health and Human Services, Office of the Surgeon General 2009.

[7] US Department of Housing and Urban Development. Healthy homes and lead hazard control: Making homes healthier for families. HUD 2008. Accessed 2013 May 20. URL: http://www.hud.gov/offices/ lead/healthyhomes/index.cfm.

[8] Centers for Disease Control and Prevention. Healthy homes: Comprehensive programs. CDC 2009. Accessed 2013 May 20. URL: http://www.cdc.gov/ HealthyHomes/ByAudience/ Programs_ Comprehensive.html.

[9] US Environmental Protection Agency. "Healthy Homes." EPA New England 2010. Accessed 2013 May 20. URL: http://www.epa.gov/region1/healthyhomes/.

[10] US Census Bureau. American Housing Survey for the United States: 2007. Accessed 2013 May 20. URL: http://www.census.gov/prod/2008pubs/h150-07.pdf.

[11] National Center for Healthy Housing. Consumer Survey Report 2009. Accessed 2013 May 20. URL: http://www.nchh.org/Home.

[12] Fisk WJ, Lei-Gomez Q, Mendell MJ. Meta-analyses of the associations of respiratory health effects with dampness and mold in homes. Indoor Air 2007;17(4):284-96.

[13] Centers for Disease Control and Prevention. Healthy Youth: Asthma. CDC 2013 May 20. Accessed 2010. URL: http://www.cdc.gov/healthyYouth/asthma/.

[14] North Carolina Rural Economic Development Center, Inc. Population databank. REDC 2009. Accessed 2013 May 20. URL: http://www.ncruralcenter.org/.

[15] McAlister AL, Perry CL, Parcel GS. How individuals, environments, and health behaviors interact: Social cognitive theory. In: Glanz K, Rimer BK, Viswanath K, eds. Health behavior and health education: Theory, research, and practice. San Francisco, CA: Jossey-Bass, 2008: 69-188.

[16] Kegler MC, Miner K. Environmental health promotion interventions: Considerations for preparation and practice. Health Educ Behav 2004;31(4):510-25.

[17] Jacobs DE. A qualitative review of housing hazard assessment protocols in the United States. Environ Res 2006;102:13-21.

[18] Healthy People 2020. Proposed Healthy People 2020 objectives. Department of Health and Human Services 2010. Accessed 2013 May 20. URL: http://www.healthypeople.gov/.
[19] Cox P. Guilford County Health Department. Personal Correspondence with Herendeen LA. 1 June 2010.
[20] Dalporto A, McKee-Huger B. Greensboro Housing Coalition. Personal Correspondence with Herendeen LA. 10 June 2010.
[21] NC Department of Environment and Natural Resources. NC strategic plan to eliminate childhood lead poisoning. North Carolina: Childhood Lead Poisoning Prevention Program, Children's Environmental Health Branch, Division of Environmental Health, Department of Environment and Natural Resources 2009.
[22] US Census Bureau. Guilford County, North Carolina. American Factfinder 2008. Accessed 2013 May 20. URL: http://factfinder.census.gov/.
[23] Guilford County Department of Public Health. Community Health Assessment. Guilford County, North Carolina, Department of Public Health 2005.
[24] Oldenburg B, Glanz K. Diffusion of Innovations. In Glanz K, Rimer BK, Viswanath K. Health behavior and health education: Theory, research, and practice. San Francisco, CA:Jossey-Bass, 2008:313-33.
[25] Jacobs DE, Kelly T, Sobolewsk, J. Linking public health, housing, and indoor environmental policy: Successes and challenges at local and federal agencies in the United States. Environ Health Perspect 2007;115(6):976-82.
[26] National Center for Healthy Housing. Housing interventions and health: A review of the evidence. Columbia, MD: National Center for Healthy Housing, 2009.

In: Environmental Health
Editors: IL Rubin and J Merrick

ISBN: 978-1-62948-155-5

Chapter IV

Home-based caregiver assessment to improve children's health

Cherish Holt, RN, BSN, Randi B Kotal-Lee, RN, BSN, Brette Winston, RN, BSN, Ashley C Deringer, RN, BSN, Janice Nodvin, BA and Maeve Howett*, PhD, APRN, CPNP-PC, IBCLC

Emory University, Nell Hodgson Woodruff School of Nursing and Institute for the Study of Disadvantage and Disability, Atlanta, Georgia, US

Abstract

Grandparents Raising and Nurturing Dependents with Disabilities (Project GRANDD) is a support program for grandparents who are the primary caregivers of grandchildren with developmental disabilities, chronic illness, behavior or learning difficulties. A critical problem contributing to diminished quality of life for the family is the grandparents' own poor health, as well as lack of access to available resources. This project pairs nursing students from the local university

* Correspondence: Maeve Howett, Associate Clinical Professor, 1520 Clifton Rd., Atlanta, GA, USA, 30322, Telephone: 404-727-8170 , E-mail: mhowett@emory.edu.

with grandparents participating in the GRANDD support program to improve the caregiving grandparent's health. Methods: Twelve nurse practitioner students were paired with six grandparents. Students performed home visits and assessments, health histories, environmental assessments and windshield surveys over the course of several weeks. Follow-up visits for medication reconciliation, referral to resources, or for patient health education were conducted in person or via phone. Results: The grandparents demographic characteristics were: African American females, on fixed incomes on or near poverty level, had multiple co-morbidities, and showed reluctance in seeking help because of their role as primary caretaker. Nursing students demonstrated increased awareness of the cycle of disability and disadvantage, and improved skills that will enable them to aid vulnerable individuals in accessing community resources. Conclusions: This pilot program had a positive impact on students and grandparents. Much is needed regarding outcome measurement and standardization of methods, but the initial results demonstrate a beneficial relationship between nurse practitioner students and Project GRANDD participants, contributing to the ongoing effort to break the cycle of disadvantage and disability in this vulnerable population.

Introduction

The number of grandparents raising their grandchildren is increasing in today's society. During the 1980s there was a forty-four percent increase in the number of children living with grandparents and in a third of these homes, both parents were absent (1). Although this phenomenon is being observed in a multitude of ethnic groups, it appears to be most pervasive in African Americans (1). By the mid-1990's, approximately thirteen and a half percent of African American children were under the sole care of a grandparent (1). This can be compared to six and a half percent of Hispanic children, and around four percent of white children (1). The custodial grandparent relationship is globally prevalent. Studies have examined this relationship in various populations and countries, including Native American and European (7,11).

There are a multitude of circumstances surrounding grandparents' assumption of parental roles and responsibilities. Parents with substance abuse problems, child neglect, parents unable to care for a child, psychological problems, child abandonment, and involvement of the child welfare department were found to be the primary reasons grandparents become primary caregivers to their grandchildren (12). Other potential causes include teenage pregnancy, illness or death of a parent, working far from home, and incarceration of one or both parents (10). These often difficult circumstances increase the children's likelihood of experiencing stress and emotional issues (2).

As the number of grandparents raising grandchildren continues to rise, it has been acknowledged that many of these grandparents are caring for grandchildren who not only have emotional disturbances related to separation from parents and or abuse and neglect, but who have developmental delays and disabilities as well (8). Many of the children have multiple health and behavioral problems. The most common disabilities are learning disabilities, developmental delays, attention deficit hyperactivity disorder, intellectual deficits, speech or hearing problems, and neurological problems (5). There is increasing evidence that childhood stress (14) has negative long-term consequences on an individual's health, potentially contributing to unhealthy pregnancies and subsequently a cycle of unhealthy children and increased family stress.

Grandparents who are providing care to their grandchildren have negative health outcomes and increased life stressors associated with and exacerbated when the children have developmental disabilities. Custody is also often informal with no established legal guardianship which makes it difficult for grandparents to access medical care and educational support services for their grandchildren (10). In addition, the mean age of custodial grandparents is 59.4 years (2), which has significant health implications as the grandparent is not eligible for Medicare services in the United States until age 65. Because of the lack of medical insurance, ineligibility for Medicare and a low income, grandparent caregivers often neglect their own health.

Grandparents who are primary caregivers report poorer health outcomes and functional limitations, more chronic condition exacerbations, comorbidity worsening, self-assessed health decline, and limitations to their activities of daily living than non-care-giving grandparents (1). Custodial grandparents actually report worse self-assessed health than non-caregiving grandparents with the most frequently reported health issues being elevated blood pressure and arthritis (15). Twenty-five percent of grandparent caregivers were found to have clinically significant depression (9), and sixty-five percent of grandparents raising grandchildren with disabilities reported a health problem such as coronary disease, orthopedic issues, uncontrolled diabetes, and ulcers (12). Grandparents raising children with physical or behavioral problems have been found to be among the least likely to seek and receive counseling and other help for themselves (3). Because of housing prices, there has been a scattering of families living in poverty out of urban areas and into the suburbs, adding to their isolation and lack of awareness of resources available to them. The increased likelihood of depression, and worsening and neglect of health among custodial grandparents increases the burden of caring for children with disabilities and was the impetus for this pilot project with Project GRANDD participants.

'Relative caregiver', or 'kinship care', can be any related family member stepping in to care for children when the biological parent is unable. Globally, kinship care is quite common after wars or displacement in refugee communities, in AIDS ravaged countries, and countries where immigration to economic opportunities require children to be left behind. In the United States it is most commonly done by grandparents for their grandchildren -- and is often an alternative to foster care placement. These grandparents are frequently the only stable presence in their grandchildren's lives (10). If the grandparent were unable to care for the grandchild with disabilities, the children would in all likelihood be split from their sibling group and have worse health outcomes related to foster care. Therefore, it is imperative for the children that the caregivers' health becomes a priority and they maintain their health through preventative health care services and chronic disease management.

Project GRANDD (Grandparents Raising and Nurturing Dependents with Disabilities) is a program which was established by the Institute for the Study of Disadvantage and Disability (ISDD) in 2005. This program offers support to grandparents who are raising grandchildren with disabilities, chronic illness, and behavior or learning difficulties. The ISDD provides an array of services to these grandparents such as support groups, educational meetings, in-home visits, case management services and referrals for medical, legal and counseling services. Currently the project serves eighty grandparent caregivers and nearly 200 grandchildren. All of the caregivers involved in Project GRANDD are African-American, female grandparents who have fixed incomes at or below the poverty level in the United States. Many are unable to work due to the complexities of the disabilities of their grandchildren.

Our study

The initial pilot involved twelve nurse practitioner students paired with six grandmothers participating in Project GRANDD and living throughout the metropolitan area surrounding Atlanta, Georgia in the southeastern United States. Faculty from the school of nursing met to design an appropriate intervention and assessment strategy consisting of an initial intensive meeting facilitated by the Project GRANDD program director who would serve as a familiar and trustworthy introduction for the family. Each student pair was assigned a grandmother client and given instructions on how to do a thorough health assessment via house visits and phone calls.

The assessment began with a "windshield" survey of the community, noting such things as access to grocery stores, pharmacies, healthcare facilities, childcare centers and transportation. Descriptive comments were documented related to amounts of trash and graffiti, fast food establishments and visible loitering or other risk factors in the neighborhood. Each student pair then made an initial visit to their respective client to gather a detailed medical and surgical history, list of current medications, conduct a thorough review of systems and perform a comprehensive physical examination of the grandmother. The students then did a room-by-room safety and environmental assessment of the home looking for fall risks, trip hazards, mold and mildew problems or other environmental risks. Finally, they asked the grandmothers to reflect on what they believed their most pressing health priorities were and how the students could help them.

Upon return to the nursing school, students met with faculty to write up their assessment findings, practice presenting a case study and to develop a prioritized care plan for the grandmother based upon their assessment findings. The Project GRANDD administrator provided information about resources in the community that could be used to meet objectives in the care plans. In addition, nursing faculty and the GRANDD administrators provided clinical guidance and debriefings for the students throughout the course of the pilot.

Findings

All participants are custodial grandparents, all raising more than one grandchild. In addition, these grandparents are all raising grandchildren with physical and/or developmental disabilities. Grandparents tended to neglect their own care due to various constraints, ranging from limited mobility, no transportation, financial burden, or fear of abandoning the needs of their grandchildren. There were several other commonalities among the six grandparent participants. All participants were unemployed African-American grandmothers. We noted that they all were on Medicaid, and those who were over age 65 were insured by Medicare. Regarding living arrangements, none of the participants owned their own home: one was living with relatives, three rented apartments, and two lived in subsidized rental housing. None of the housing was ideal for the families and often lacking in adequate space, plumbing or safety features. These grandparents also reported multiple severe co-morbidities, such as cardiac disease, hypertension, depression and arthritis..

The NP students gained an awareness of the isolation and poverty that these grandparents suffer in and how this affects their health. They became acutely aware of how the health of the grandparent has the potential to influence the children's health, housing, education and custodial arrangements. They were also introduced to many resources for the grandparents that they were not aware of initially, such as the Atlanta Regional Commission, which provides a referral hub for utility bill assistance, housing referrals and legal aid. Furthermore, students performed searches for other resources their grandparents might find useful. For example, one grandparent was not aware of free or discounted shuttles operated by the local transit authority, county libraries, or churches. Students were made aware of community resources (or lack thereof), and spent time integrating appropriate resources for their clients.

Discussion

The overall intent of this pilot program was to evaluate the healthcare needs of custodial grandparents and their home environments, and expose nurse practitioner students to this vulnerable population and to raise awareness of the concept of health disparities. Students responded positively to the project and we anticipate more nurse practitioner students will show an interest in the program in the future. To objectively assess the impact of this pilot project, qualitative measures will need to be used in subsequent projects. Grandparents also gave universally positive comments and feedback concerning the students' involvement, but no scales or questionnaires were used to evaluate this impact. Continuation of this project would have two main outcomes: refining the evaluation piece would yield more data, and the project would enhance the practicum experiences for students. The benefit of this project is three-fold: grandparents have a unique opportunity to interact one-on-one with a healthcare provider for an extended period of time, they receive a health assessment with appropriate referrals to improve their physical well-being, and students have an opportunity to interact with a vulnerable population that will affect their maturation as professionals and ability to advocate for patients.

Therefore, despite the lack of quantitative evidence in this pilot program, we do believe it is a very satisfactory start to what could become a long-term relationship between nursing students and Project GRANDD participants.

We primarily anticipated that custodial grandparents would exhibit better health literacy and more self-care behaviors. All of the grandparents were in relatively poor health and dealing with multiple life stressors that exacerbated their health status. Anecdotal evidence did demonstrate that the grandparents enjoyed the relationship with their respective nurse practitioner students. In some cases there was verbalization of improved health literacy or decisions to engage in more self-care. Medication reconciliation and education was universally needed by the grandparents and several needed encouragement to make specialist visits with providers for surgery or follow-up from an earlier health crisis. At least one grandparent made a very important decision to undergo surgery that was needed for a long time; while for many months this individual was reluctant to do so based on her fear of abandoning the grandchildren and not providing them with proper care in her absence. Additionally, information on their medications, diagnoses (and how to deal with these on a day to day basis), stress relief and healthy diets were provided to the participants. It is unknown if this guidance was helpful to the grandparents because we did not measure stress levels or diet before and after the program, but all the participants reflected that the visits had a positive impact simply from someone spending time with them and asking about their health and needs, as so often the focus of health providers was on their grandchildren's needs.

It was also anticipated that the students would display increased awareness of the cycle of disability and disadvantage, with ideas or knowledge that will help interrupt the cycle in the future. This was not the students' first experience in community health with underserved populations, but it was their first experience with the unique situation of older adults caring for grandchildren with disabilities. Learning to do home visits increased their ability to care for this underserved population and because they performed the assessment in the home, the grandparents felt more comfortable and spoke openly and freely about their health history and concerns. The students could see firsthand the safety risks and neighborhood factors that affect health. We therefore have no doubt that students are more aware of this cycle after participating in the pilot program. Furthermore, the integration of community resources and educational needs specific to clients is a valuable skill for nurses and nurse practitioners. In particular, the ability to locate resources and enable clients to access them is crucial in community health and underserved populations. Students were therefore more informed and able to navigate various community resources after participation in this program, a skill that we believe is imperative to breaking the cycle of disadvantage.

Due to the exploratory nature of this pilot program, we did not use any objective or standardized measures for the anticipated outcomes. We therefore cannot say with absolute certainty that we positively impacted the grandparents' health or student perceptions, so for the purposes of this paper we must rely on anecdotal and qualitative evidence. This is an incentive to proceed with a more data driven plan in order to objectively demonstrate the positive impact, even some improvement in health literacy or changes in attitude and healthcare decisions of the participants. The challenge will lie in maintaining the quality of student-grandparent relationships in the midst of data collection, as some may find that numerous evaluations detract from the positive interactions or disrupt the care environment (6). There were also various time and travel constraints, which led to inconsistency among student pairs, so that some students visited their clients multiple times and others only once or twice. There may be a natural inclination for student-grandparent pairs who interacted multiple times to demonstrate more positive outcomes, which could lead to bias.

It will also be important to develop specific long-term goals for grandparents and students over the next few years. This will require even more effort for all parties involved, but may help with transitioning subsequent student groups into leadership roles within the program. Building a long-term relationship will take time, but we believe it would be worthwhile.

Acknowledgments

The authors of this paper thank Leslie Rubin and Darla Ura for their input, guidance, and encouragement throughout this pilot program. This chapter is a revised and adapted version of an earlier publication: Holt C, Kotal-Lee RB, Winston B, Deringer AC, Nodvin J, Howett M. The use of home-based caregiver assessment to improve children's health: A pilot project. Int J Disabil Hum Dev 2013;12(4).

References

[1] Fuller-Thompson E, Minkler M. African American grandparents raising grandchildren: a national profile of demographic and health characteristics. Health Soc Work 2000;25(2):109-18.

[2] Ghuman HS, Weist MD, Shaefer ME. Demographic and clinical characteristics of emotionally disturbed children being raised by grandparents. Psychiatr Serv 1999;50(11):1496-98.

[3] Hayslip B, Shore J, Henderson CE, Lambert PL. Custodial grandparenting and the impact of grandchildren with problems on role satisfaction and role meaning. J Gerontol 1998;53(3):S164-73.

[4] Hudson C. (n.d.). ELEV8-Crying Child. Health advice & inspiration for black America. Accessed 2013 May 20. URL: http://elev8.com/spirit/relationships-daily-offering/cherylhudson/parents-in-blended-families-stop-with-all-the-drama/

[5] Janicki MP, McCallion P, Grant-Griffin L, Kolomer SR. Characteristics of the grandparents and the children with disabilities for whom they care. J Gerontol Soc Work 2000;33:33-35.

[6] Krout J, Bergman E, Bianconi P, Caldwell K, Dorsey J, Durnford S, et al. Intergenerational service learning with elders: multidisciplinary activities and outcomes. Gerontol Geriatr Educ 2010;31:55-74.

[7] Letiecq BL, Bailey SJ, Kurtz MA. Depression among rural native American and European American grandparents rearing their grandchildren. J Fam Issues 2008;29:334-56.

[8] McCallion P, Janicki MP, Kolomer SR. Controlled evaluation of support groups for grandparent caregivers of children with developmental disabilities and delays. Am J Ment Retard 2004;109:352.

[9] Minkler M, Fuller-Tomson E, Miller D, Driver D. Depression in grandparents raising grandchildren: results of a national longitudinal study. Arch Fam Med 1997;6:445-52.

[10] Neely-Barnes SL, Dia DA. Families of children with disabilities: a review of literature and recommendations for interventions. J Early Intensive Behav Interv 2008;5(3):93-07.

[11] Today's research on aging: the health and well-being of grandparents caring for grandchildren. Population Reference Bureau 2011. Accessed 2013 May 20. URL: http://www.prb.org/pdf11/ todaysresearchaging23.pdf

[12] Sands RG, Goldberg-Glen RS. The impact of employment and serious illness on grandmothers who are raising their grandchildren. J Women Aging 1998;10(3):41-58.

[13] Sangweni Y. Raising grandchildren affects health of grandparents. Essence.com 2010. Accessed 2013 May 20. URL: http://www.essence.com/2010/03/06/raising-grandchildren-can-affect-health/

[14] Shonkoff JP, Garner AS, Siegel BS, Dobbins MI, Earls MF, McGuinn L, et al. The lifelong effects of early childhood adversity and toxic stress. Pediatrics 2012;129(1): e232-e46.

[15] Smithgall C, Mason S, Michels L, LiCalsi C, George RM. Caring for their children's children: assessing the mental health needs and service experiences of grandparent caregiver families. Chapin Hall University of Chicago 2006. Accessed 2013 May 20. URL: http://www.chapinhall.org/ research/report/caring-their-childrens-children.

In: Environmental Health
Editors: IL Rubin and J Merrick
ISBN: 978-1-62948-155-5

Chapter V

Effect of health problems on the quality of life in children and young adults with spina bifida

David Wood*, MD, MPH[1,2], Graham Watts, PhD[2], Kimberlea Hauser, MBA[3], Panta Rouhani, MPH[4] and Jaime Frias, MD[4]

[1]Department of Medicine, University of Florida, Gainsville, Florida, US
[2]Institute for Health, Policy and Evaluation Research, Duval County Health Department, Jacksonville, Florida, US
[3]Birth Defects Center, University of South Florida, Tampa, Florida, US
[4]Department of Medicine, University of South Florida, Tampa, Florida, US

Abstract

In this chapter we assessed the impact of health problems in children and young adults with spina bifida on their health-related quality of life. Design: We surveyed 124 children and young adults with spina bifida in central Florida identified through multiple sources including multi-specialty clinic rosters, Title V clinic registries, and community-based

* Correspondence: Professor David Wood, MD, MPH, UF College of Medicine-Jacksonville, Jacksonville Health and Transition Services (JaxHATS) and Center for Health Equity and Quality Research, 580 W 8th St, Room 7005, Mailstop T-60, Jacksonville, FL 32209 United States. Tel: 904-244-9233 ; Fax: 904-244-9234 ; E-mail: david.wood@jax.ufl.edu.

spina bifida associations. The Child Health Questionnaire, (CHQ) and SF-36 were used to assess health related quality of life (HRQOL). Results: The sample included 84 children aged 5 to 17 years old and 40 young adults age 18 to 30 years old. In the past year, more than half (55%) had experienced a bladder infection, 40% pressure sores, 39% an allergy to latex, 38% scoliosis or kyphosis, 32% chronic pain, and 27% a weight problem. Children with chronic pain (versus those without chronic pain) reported the largest difference in HRQOL scores in both the psychosocial or physical health areas: physical HRQOL: 24.7 vs. 37.5, $p < .01$; psychosocial HRQOL: 48.7 vs.41.4, $p < .01$. Adults with chronic pain reported significantly lower psychosocial health scores (46.0 vs. 58.1, $p < .01$). Other conditions associated with significantly reduced HRQOL scores were pressure sores and bladder infections in children and latex allergy in adults. Conclusions: In both children and adults with spina bifida, chronic pain has a substantial negative impact on physical and psychosocial health-related quality of life. In order to maximize the quality of life of persons with spina bifida, increased efforts are warranted to identify and control chronic pain.

Introduction

Advances in surgical and medical treatments have helped to dramatically increase the survival of individuals with spina bifida (SB) or meningomyelocele into adolescence and adulthood (1). However, persons with SB experience serious health problems that compromise their quality of life and function. Approximately 85% have hydrocephalus necessitating the placement of ventriculo-peritoneal (VP) shunts. VP, while greatly reducing mortality and significant morbidity, frequently become infected and can cause significant cognitive impairment. Virtually all persons with SB experience neurogenic bladder to some degree, which is often complicated by recurrent urinary track infections, vesico-ureteral reflux and renal damage. Insensate skin below the level of the spinal lesion is susceptible to breakdown and pressure ulcers, requiring frequent medical visits or lengthy hospitalizations (2). Severe IgE mediated hypersensitivity to latex has been reported to be present in up to 47% of persons with SB. Progressive kyphosis and scoliosis is seen in 10-15% of persons with SB and can impact on mobility, proper seating, and respiratory function (4,5). The frequency, duration, or severity of pain have not been well studied in persons with SB, but usually are reported to be due to multiple potential sources, including some of the conditions cited above (6).

Chronic medical problems can significantly compromise the social, emotional and physical well being of persons living with SB (7,8). With the increase in life expectancy, scientists and clinicians are directing more attention on improving the quality of life and function of persons with SB. Moreover, it is important for clinicians to understand the relationship between these health problems and health related quality of life (HRQOL) in order to provide care that maximizes the health and functional capacities of patients and their families (7,9).

There is, however, a lack of information and understanding on the relationships among SB, its medical complications, and HRQOL. Given the varying perspectives of patients and their providers, this dearth of knowledge may lead to mismatched priorities between patients and their treating physicians. Several factors contribute to the paucity of HRQOL data. First, health related physicians or other providers do not typically assess quality of life. Second, reliable measurement of HRQOL requires the use of validated questionnaires which clinicians are not accustomed to using. Lastly, HRQOL is a complex set of constructs that have complicated relationships with clinical disease states (10). The purpose of this study, therefore, is to describe the health-related quality of life associated with specific health problems of persons with SB using validated HRQOL instruments.

Our working hypotheses were: 1) chronic and potentially disabling health conditions are common in children and young adults with Spina Bifida and 2) these conditions affect, and potentially reduce, their physical, social and emotional quality of life. More specifically, we were interested in investigating the relationships between the different domains of quality of life and the following health problems: chronic pain, urinary tract infections, skin breakdown, scoliosis/kyphosis, obesity or latex allergy.

Our study

The study was funded by the Florida's Developmental Disabilities Council (FDDC) and reviewed by the Institutional Review Boards of the University of South Florida (USF) and Arnold Palmer's Hospital for Women and Children.

As part of the study we sought to develop a universal registry of persons with SB in Central Florida (Tampa/St. Petersburg, Orlando and surrounding areas). We identified persons with SB through the two multi-disciplinary, hospital-based SB clinics (Shriners and Arnold Palmer hospitals) in the area Title V programs covering Central Florida, and through print, voice, and visual advertising with the Spina Bifida Associations, the Florida Youth and Transition listserv, and the Developmental Disabilities Program. We were unable to develop the registry because the hospital-based clinics could not release identifying information on patients. We did ask them to contact patients with SB and their families and r recruit them to participate and obtain their verbal consent for us to contact them. Once they agreed to participate, caregivers of children younger than 18 years old and respondents 18 years and older were provided written informed consent. Individuals received a self-addressed, stamped return envelope, a cover letter explaining the study, an informed consent letter, which asked for voluntary participation in the study, and a contact information sheet. Participant recruitment continued for a 19 months period. The final sample was composed of 64 respondents from clinic rosters and 60 from non-clinical settings. No significant differences were found between the clinic sample and the non-clinic sample. No information was available on the patients or families who refused to participate when contacted by the Care Coordinators in the hospital-based clinics and in the Title V program.

Instruments

Separate health related quality of life batteries were used for children and young adults. For children aged 5 to 17, we used the Child Health Questionnaire (CHQ) version, which has been validated in multiple studies with different health conditions and shows excellent internal reliability, and good responsiveness to variations in health status (11-13). The CHQ has 14 basic content domains/scales that aggregate to physical, and psychosocial summary scales (14). These HRQOL summary scores for physical and psychosocial health are computed from six subscales. The physical HRQOL includes respondent ratings of physical functioning, role/social-physical, general health, bodily pain, and family impact from physical limitations. The psychosocial HRQOL score includes respondent ratings of the impact on social role function due to social-emotional or behavioral limitations, ratings of self-esteem, mental health states, and behavior problems. A sample item from the CHQ asks: “During the past 4 weeks, has <child’s name> been limited in doing things that take a lot of physical energy, such as playing sports, or engaging in strenuous activity (swimming, wheelchair sports, propelling a wheelchair for a long distance or time period) due to health problems?" Responses include: a) Yes, limited a lot, b) Yes, limited some, c) Yes, limited a little, d) No, not limited, and e) Refuse.

In young adult respondents we used the SF-36® to measure HRQOL, the most widely used and validated quality of life measurement tool (15). The SF-36 contains 36 questions that are also summarized into physical health and psychosocial health summary scales. The physical health scale (or Physical Component Scale) includes respondent ratings of physical functioning, role physical, bodily pain, and general health. The psychosocial health scale (or Mental Component Scale) includes respondent ratings of the following areas: vitality, role emotional, mental health, and social functioning. A sample item asks: "Does your health now limit you in performing vigorous activities such as participating in wheelchair sports, swimming, propelling a wheelchair for a long distance or time period, moving heavy objects?" Responses include: a) Yes, limited a lot, b) Yes, limited a little, c) No, not limited at all, and d) Refuse. The CHQ and SF-36 are very comparable in the domains of physical, social and mental health functioning because the CHQ was based on the constructs developed and tested in the SF-36.

Demographic questions covered the child's and primary respondent's age, gender, family living situation, and urban or rural residence as determined by the zip code of the primary residence. The family's living situation was assessed through a series of questions that inquired about family and household composition and was summarized as living with both parents (two-parent family), living with a single parent, or living alone. The respondents were asked if they experienced a list of health problems during the past year including; hydrocephalus, bladder infection, pressure sores, latex allergy, scoliosis, weight problem, chronic pain, and shunt revisions (at least once in the past five years). The responses available to these were; No, Yes, Don't Know.

Procedures

Administration of the survey was interviewer guided. The parents or primary caretaker of subjects less than 18 years of age completed the surveys. Subjects 18 years of age or over completed the surveys themselves. Respondents over 18 were given the option of having their caretaker respond to the interview. However, all persons 18 years and over chose to respond for themselves. One week before the interview, the respondents received a packet that contained the Likert-scale response sets to the different questions. Interviews were either conducted by one of the authors (PR) or by interviewers trained and supervised by her. The interviews varied in length from 45 to 90 minutes, depending upon the degree to with the respondent chose to give information on the open-ended questions.

Analysis

The Statistical Package for the Social Sciences (SPSS), Windows version 11.5, was used for analysis. Basic descriptive statistics and frequencies were computed for all sociodemographic and health status conditions as appropriate. CHQ and SF-36 scores were transformed to 0-100 scales. One sample t-tests compared CHQ and SF-36 mean composite scores to determine whether HRQOL in the study sample differed significantly from established population values published in The Child Health Questionnaire (16) and SF-36 Health Survey manuals (15). We used independent sample t-tests to compare mean CHQ and SF-36 composite scores of those with or without specific health conditions listed above. In order to avoid a Type I error, we set the significance level for a rejecting the null hypothesis at $p < 0.01$ rather than $p < 0.05$. Due to the small sample size we could not do a full Bonferonni correction (which is calculated as $p < 0.05$/number of comparisons (26, see table 4) or 0.05/26=0.0019).

Findings

Socio-demographic data are presented in table 1. Slightly more than one-third (36%) of the respondents were between 5 and 11 years old, 32% were between 12 and 17 years old, and 32% were between 18 and 30 years old. Approximately equal proportions of both genders (female 52%) were included in the sample. The majority of respondents lived in a two-parent family (63%), a minority lived in a single parent household (29%), and a few lived alone. More than half (57%) resided in urban centers. We had information on the level of the lesion for 91 respondents; in over two-thirds (68%) the lesion was located in the lumbar region of the spine, with 15% having their lesion in the sacral and lumbosacral region, and 16% had lesions in the thoracic or cervical regions.

Table 1. Selected socio-demographic characteristics of respondents

Socio-demographics	% (n)
Age in years	
5-11	36% (45)
12-17	32% (39)
18 to 30	32% (40)
Gender	
Male	48% (59)
Female	52% (65)
Living situation	
Family	63% (78)
Single Parent	29% (36)
Alone	8% (10)
Type Survey	
Individual	27% (34)
Primary Care Giver	73% (90)
Residence	
Rural	43% (54)
Urban	57% (70)
Level of the lesion	
Cervical	01% (01)
Thoracic	15% (14)
Lumbar	68% (62)
Lumbar/Sacral	15% (14)

Percentages may not sum to 100% due to missing data. Sample size was 124.

Table 2. Self reported health status and conditions

Conditions	(n) %
Hydrocephalus	65% (81)
Shunt revision in past five years	30% (38)
Bladder infection	55% (68)
Skin breakdown/pressure sores	40% (50)
Latex allergies	39% (48)
Scoliosis/Kyphosis	38% (47)
Chronic Pain	32% (39)
Weight problem	27% (34)

Sample size was 124.

In table 2 we report the responses to questions related to the presence of specific health problems associated with SB. Sixty-five percent (n=81) reported having hydrocephalus and almost half of these (30% of total population, n = 38) of these had at least one shunt revision in the last five years. In the past year, more than half (55%) had experienced a bladder infection, 40% pressure sores, 39% an allergy to latex, 38% scoliosis or kyphosis, 32% chronic pain, and 27% a weight problem. We found that 84% of the individuals in our sample had experienced one or more health problems.

CHQ and SF-36 physical and psychosocial composite health scores are presented in table 3. Each composite score had a theoretical range from 0 to 100, with higher scores indicating a better reported health related quality of life. The CHQ physical population mean, as determined by the responses of thousands of well children, is 53.2 while the mean for our sample of children with spina bifida was significantly lower at 33.90±1.8 ($p < 0.01$). The CHQ psychosocial health population mean is 51.1 whereas the mean for our children with spina bifida was significantly lower at 46.67±1.9 ($p < 0.01$). For the adults, the SF-36 physical health composite population mean was 53.8 and the mean for our sample of young adults with spina bifida was significantly lower at 45.87 ± 1.0 ($p < 0.01$), indicating lower overall physical health-related quality of life. The mean population SF-36 psychosocial health score was 48.0, and the psychosocial HRQOL score for our sample was significantly higher at 53.67 ±1.95 ($p < 0.01$), indicating a higher psychosocial health-related quality of life for our sample compared to the general population.

Table 3. CHQ and SF-36 composite health and functioning scores

Composite Health Scores	General Population Mean± s.e	Study Sample Mean ± s.e (n)
CHQ Physical	53.2	33.90± 1.77 (87)*
CHQ Psychosocial	51.1	46.67± 1.28 (87) *
SF-36 Physical	53.8	45.87±1.55 (34) *
SF-36 Psychosocial	48.0	53.47±1.88 (34)*

Sample size was 124, scores range from 0 to 100 and higher scores indicated better health state and related quality of life.
*indicates p < .01.

Table 4. Comparison of physical health and psychosocial health scores those with versus without specific health problems

Health Condition	CHQ Physical Health Mean ± s.e. (n)	CHQ Psychosocial Health Mean ± s.e. (n)	Adult SF-36 Physical Health Mean ± s.e. (n)	Adult SF-36 Psychosocial Health Mean ± s.e. (n)
Hydrocephalus				
No	37.6 ± 4.4 (19)	44.32 ± 3.27 (19)	--	--
Yes	31.7 ± 2.1 (56)	46.81 ± 1.54 (56)	45.84 ± 1.94 (25)	52.06 ± 2.3 (25)
Shunt revision last 5 years				
No	33.72 ± 3.6 (23)	46.97 ± 1.88 (23)	46.12 ± 2.22 (14)	55.34 ± 3.0 (14)
Yes	30.16 ± 2.74 (28)	48.30 ± 2.41 (28)	43.06 ± 3.86 (9)	46.78 ± 4.1 (9)
Bladder infections				
No	**40.41 ± 2.83 (41)**	46.93 ± 1.81 (41)	47.21 ± 2.70 (13)	**57.76 ± 1.4 (13)**
Yes	**28.10 ± 1.86 (46)**	46.44 ± 1.83 (46)	45.03 ± 1.89 (21)	**50.81 ± 2.78 (21)**
Chronic Pain				
No	**37.48 ± 1.93 (61)**	**48.67 ± 1.43 (61)**	48.12 ± 1.77 (21)	**58.10 ± 1.37 (21)**
Yes	**24.73 ± 3.32 (25)**	**41.39 ± 2.5 (25)**	42.23 ± 2.65 (13)	**45.97 ± 3.56 (13)**
Pressure sores				
No	**36.59 ± 2.2 (56)**	48.49 ± 1.44 (56)	44.24 ± 2.3 (17)	53.08 ± 2.77 (17)
Yes	**29.03 ± 2.85 (31)**	43.40 ± 2.41 (31)	47.49 ± 2.1 (17)	53.84 ± 2.62 (17)
Latex allergy				
No	36.22 ± 2.2 (53)	47.54 ± 1.64 (53)	46.62 ± 1.88 (18)	**57.44 ± 1.44 (18)**
Yes	30.33 ± 3.04 (32)	45.83 ± 2.14 (32)	45.02 ± 10.29, 16)	**48.99 ± 3.17 (16)**
Weight problem				
No	34.06 ± 2 .11 (63)	48.32 ± 1.43 (63)	47.45 ± 1.85 (24)	55.41± 1.97 (24)
Yes	32.96 ± 3.46 (23)	41.71 ± 2.6 (23)	42.07 ± 8.13, 10)	48.79 ± 2.63 (10)

Scores ranged from 0 to 100 and higher scores indicated better health state and related quality of life.
Bolded font shows where differences exist with a p value of < 0.01
--- = values not posted because n is too small (n = 3) for valid comparisons.

Table 4 presents differences in CHQ and SF-36 physical and psychosocial scores for persons with and without specific health problems. Compared to children not having the reported health problems, physical health scores were significantly lower for children reporting bladder infection ($p < .01$), chronic pain ($p < .01$), and pressure sores ($p < .01$). On the psychosocial scale, children with chronic pain had a significantly lower mean CHQ psychosocial health composite score ($p < .01$) as compared to children who did not report chronic pain. Young adults with bladder infection ($p < .01$), chronic pain ($p < .01$), and latex allergies ($p < .01$) reported lower mean psychosocial health scores. There were no significant differences in physical health scores for young adults across all reported health conditions.

Discussion

This study assessed the relationship between reported health problems and health-related quality of life (HRQOL) of children and young adults with SB in central Florida. Consistent with other studies among persons with spina bifida, we found that over 80% had one or more significant health problems. In the past year, 55% had a history of urinary track infection(s), 40% had pressure ulcers, 27% had weight problems, 39% had latex allergy, and 32% had chronic pain. Previous national and international studies have found similarly high prevalence of these health problems in people with SB. In a study done in Cologne, Germany, 41% of persons with SB had latex sensitivity (17) and in a clinical study from Chicago, 30% reported chronic pain (8).

As a result of improved medical management of the potentially fatal complications of SB, 85% of those with SB survive into adulthood compared to only 10% in the 1960s (18). Since high survival rates have been achieved, the current focus is on improving the effective management of the chronic health conditions and increasing the quality of life and function of persons with SB. HRQOL scores of our sample of children with SB were much lower than those found in the general child population. The children and young adults reported substantially lower physical health functioning, indicating a fairly significant level of disability. While this is understandable given the physical limitations resulting from SB and other co-morbid conditions, proper supports and aids can greatly increase reported physical functioning. Lower physical health status was associated with a reported history of bladder infections, pressure sores, and chronic pain. Lower psychosocial health-related quality of life was associated only with chronic pain. These findings corroborate previous research, which show that persons with SB suffer from a number of health problems that can adversely impact HRQOL (19)..

Importantly, health problems including hydrocephalus, shunt revision, urinary tract infections, pressure sores, weight problem and latex allergy, were not associated with lower psychosocial functioning in children, indicating that factors other than physical health problems are important determinants of health related quality of life. Studies have demonstrated that parents who instill in their children a sense of hopefulness, self-worth and self-esteem can mitigate the negative impact of physical disabilities on mental and social health (7,20). For example, parental hope has been positively associated with children's HRQOL independent of level of disability (5). Hope itself may mediate improved mental and social health by preventing isolation, depression, anxiety or other feelings from developing in the caretaker and child.

In our study children with chronic pain had substantially lower psychosocial and physical HRQOL. Moreover, chronic pain was associated with a larger reduction in HRQOL than were the other reported health conditions. Persons with SB have many potential sources of chronic pain, including musculoskeletal pain, peripheral neuropathies, headache and abdominal pain from chronic constipation or urinary track infections. Chronic pain can lead to increased physical and emotional disability, anxiety, sleep disturbance, school absence and lower school performance, and social withdrawal (21,22). Control of pain and other co-morbid conditions, if addressed, can improve the quality of life. Unfortunately, in our survey we found that of those reporting chronic pain, only 54% reported having seen a health provider for their pain in the past year. Therefore, approximately half of the children with chronic pain had not active medical management of their pain.

Over the past two decades increased attention has been paid to the diagnosis and management of pain in children. The focus of the research has been on specific sources of pain, approaches to pain management in children, and the impact of pain and its treatment on children's quality of life and function (23). The measurement of pain in children has been greatly enhanced by the development and validation of self-report scales, including the 'faces pain scale,' which has been used extensively in hospital care. In their assessment of the quality of inpatient care JCAHO now considers regular pain measurement as important as the consistent measurement of other vital signs. As a result, pain has been called the "5th vital sign."

In the outpatient setting, however, pain and its management have not received the same level of attention. Chronic pain is still frequently under-diagnosed in children with chronic disease and therefore not addressed (24). This probably explains why only half of our sample of children with chronic pain was seeing a physician for the pain. Even when pain is recognized, research demonstrates that pain management is often inadequate.

Our study shows that chronic pain is highly prevalent in the SB population. More attention should be paid to pain measurement and the management of pain in children and young adults. Improved detection and control of pain has the potential to increase the quality of life of persons with spina bifida.

There are a number of limitations to our study. First, the sample may not be fully representative the population of persons with SB in Florida. As detailed above, we recruited patients from a number of settings. We cannot calculate the sample response rate because the agencies and specialty clinics did not divulge the number of persons contacted in order to achieve the number of responses we ultimately received. In addition, a few of the respondents were recruited passively, by responding to an advertisement for the study in newsletters of the SB associations, and were not part of the original sample. Therefore, the potential for a response bias exists. It is possible persons with health problems were more likely to respond, which would influence both our estimates of prevalence of health conditions and also the relationship between health problems and HRQOL. However, the rates of health problems we found in our survey were very comparable to surveys from other locations and studies as well as rates from and published clinical texts (3,7,8,17-19). Furthermore, the sample size of this study is as large or larger than almost all previous US studies of health problems in persons with SB. Therefore, we believe our main results, the impact of specific health problems on HRQOL, to be valid.

Conclusion

We found that children and young adults with spina bifida frequently reported physical health problems such as bladder infections, pressure sores, and chronic pain. Among the different conditions, chronic pain had the strongest negative impact on HRQOL. Chronic pain was associated with significant decreases in both psychosocial and physical health status. The literature indicates that chronic pain in children is not adequately diagnosed nor managed. Chronic pain in persons with spina bifida, if diagnosed and managed, is likely to improve their quality of life and function.

References

[1] Buran FC, McDaniel MA, Brei JT. Needs assessment in a spina bifida program: A comparison of the perceptions by adolescents with spina bifida and their parents. Clinical Nurse Special 2002;16(5): 256-62.

[2] Ayyangar R. Health maintenance and management in childhood disability. Phy Med Rehabil Clin North Am 2002;13(4):793-21.

[3] Nickel RE. Meningomyelocele and related neural tube defects. In: Nickel RE, Desch LW, eds. The physician's guide to caring for children with disabilities and chronic conditions. Baltimore, MD: Paul H Brooks, 2000.

[4] Brown PJ. Orthopedic care of children with spina bifida: You've come a long way, baby. Orthopedic Nursing 2001;20(4):51-8.

[5] Kirpalani MH, Parkin CP, Willian RA, Fehlings LD, Rosenbaum LP, King D, Nie JA. Quality of life in spina bifida: Importance of parental hope. Arch Dis Child 2000;83:293-7.

[6] Barker E, Saulino M, Caristo MA. A new age for childhood diseases. RN 2002;65(12):33-9.

[7] Padua L, Rendeli C, Rabini A, Girardi E, Tonali P, Salvaggio E. Health-related quality of life and disability in young patients with Spina Bifida. Arch Phys Med Rehabil 2002;83:1384-8.

[8] Selber P, Dias L. Sacral-level myelomeningocele: Long-term outcome in adults. J Pediatr Orthopaedics 1998;18:423-7.

[9] Schmidt JL, Garratt MA, Fitzpatrick R. Child/parent-assessed population health outcome measures: a structured review. Child Health Dev 2002;28(3):227-37.

[10] Shang C. Emerging paradigms in mind-body medicine. J Alternative Complementary Med 2001;7(1):83-91.

[11] Alonso ME, Neighbors K, Mattson C, Sweet E, Ruch-Ross H, Berry C, Sinacore J. Functional outcomes of pediatric liver transplantation. J Pediatr Gastroenterol Nutr 2003;37:155-60.

[12] Cameron JF, Smidtst D, Hesketh K, Wake M, Northam AE. Early detection of emotional land behavioral problems in children with diabetes: The validity of the child health questionnaire as a screening instrument. Diabetic Med 2003;20:646-50.

[13] Walters E, Salmon L, Wake M, Hesketh K, Wright M. The child health questionnaire in Australia: Reliability, validity and population means. Aust NZ J Public Health 2000;24(2):207-10.

[14] Schneider WJ, Gutierrez LA, Gaebler-Spira JD. Health-related quality of life and functional outcome measures for children with cerebral palsy. Dev Med Child Neurol 2001;43:601-8.

[15] Ware JE, Kosinski M. SF-36. Physical and mental health summary scales: A manual for users of version 1, 2nd ed. Lincoln, RI: Quality Metric, 2001.

[16] Landgraf JM, Abetz L, Ware JE. The CHQ user's manual. Second printing. Boston, MA: HealthAct, 1999.

[17] Cremer R, Hoppe A, Korsch E, Kleine-Diepenbruck U, Bläker F. Natural rubber latex allergy: Prevalence and risk factors in patients with spina bifida compared with atopic children and controls. Eur J Pediatr Surg 1998;157:13-6.

[18] Bomalaski MD, Teague JL, Brooks B. The long-term impact of urological management on the quality of life of children with spina bifida. J Urol 1995;154(2 Pt 2):778-81.

[19] Cate IMP, Kennedy C, Stevenson J. Disability and quality of life in spina bifida and hydrocephalus. Devel Med Child Neuro 2002;44:317-22.

[20] Michom PE, Ellis NC, Appleton PL, et al. Impact of functional severity on self-concept in young people with spina bifida. Arch Dis Child 1995;73:48-52.
[21] Perquin CW, Hazebroek-Kampschreur AAJM, Hunfeld JAM, van Suijlekom-Smit LWA, Passchier J, van der Wouden JC. Chronic pain among children and adolescents: Physician consultation and medication use. Clin J Pain 2001;16:229-35.
[22] Sawin JK, Brei JT, Buran FC, Fastenau SP. Factors associated with quality of life in adolescents with Spina Bifida. J Holistic Nurs 2002;20(3):279-304.
[23] Howard RF. Current status of pain management in children. JAMA 2003;290:2464-9.
[24] Eccleston C, Malleson P. Managing chronic pain in children and adolescents. We need to address the embarrassing lack of data for this common problem. BMJ 2003;326:1408-9.

In: Environmental Health
Editors: IL Rubin and J Merrick

ISBN: 978-1-62948-155-5

Chapter VI

Service dogs

Yvonne W Fry-Johnson, MD, MSCR[1], Suzanne Powell, MPH[2] and Donnie Kanter Winokur[3,*]

[1]Morehouse School of Medicine, Department of Pediatrics, Atlanta, Georgia,
[2]Morehouse School of Medicine, Southeast Regional Training Center on Fetal Alcohol Syndrome, Atlanta, Georgia and
[3]National Organization of Fetal Alcohol Syndrome, Georgia Affiliate, Atlanta, Georgia, US

Abstract

Children with intellectual or behavioral disabilities have numerous challenges in their lives, including navigating their world without injury, developing appropriate interpersonal interactions, and calming of anxiety and fear. This case report is to raise awareness among an audience of individuals focused on the study, care, and management of persons with intellectual and developmental disabilities of the availability of service dogs to assist children with diagnoses including, but not limited to, fetal alcohol spectrum disorders (FASD) and autism. We are presenting findings from an informal observation of the first specifically trained

* Correspondence: Donnie Kanter Winokur, FASDempowered, Educator, 305 Todwick Drive, Roswell, GA 30075, United States. E-mail: thechancerchronicles@gmail.com. or fasdempowered@gmail.com.

service dog for the assistance of a child affected by fetal alcohol spectrum disorder (FASD). Method: We interviewed a family who has a child with FASD to collect information on the impact of the addition of the specially trained canine. Our target family relayed observations on the behavioral interactions at both the home and external environment of the child in the six months before receipt of the dog, at the time of receipt and training, and three months post incorporation into the family. We also sought information from various providers of service dogs on their uses of these assistance animals. Result: Perceived improvement in the family's ability to interact with their child, and the child's ability to interact with his environment. Conclusion: Service dogs have been shown to help individuals with a multitude of medical and mental health challenges. This case report suggests their use be considered for or offered to children with FASD, a spectrum consisting of multiple intellectual and behavioral disabilities. When an intervention like this is available to improve the quality of life of affected children and their families, the model needs to be replicated, and made available across socioeconomic lines.

Introduction

Fetal alcohol spectrum disorders (FASD) is an umbrella term for the conditions resulting from alcohol-effected pregnancies (AEP) (1). The resulting conditions, fetal alcohol syndrome (FAS), alcohol-related birth defects (ARND) and alcohol related neurodevelopmental disorder (ARND) serve as the leading preventable cause of intellectual disabilities in the world (2,3). The spectrum of disorders resulting from alcohol exposure in pregnancy impacts an inordinate amount of functioning in affected individuals (growth, organ systems, behavior patterns, communication skills, social skills, deficits in reasoning, verbal and spatial learning and memory, executive functioning, decision making, school success, job and living skills, balance and processing issues, etc), and consequently impacts their family, school, and community. When an individual is diagnosed with an FASD, anywhere along the age continuum, the diagnosis has significant implications for the child and family. The practicalities and process of taking care of children with FASD far exceeds the norm in terms of complexity and longevity, and extensively involves the healthcare, educational, and frequently but unfortunately, the judicial systems. Due to lack of clinician awareness of diagnosis and management of FASD and/or lack of skill and confidence in making the diagnosis, providers sub-optimally provide an appropriate and timely confirmation of FASD conditions (FAS, ARND, ARBD) (4,5). As a result, clinicians truly drop the ball in supporting families into and through the resultant troublesome elementary school, preteen, teen, and young adult years...where the family is left to bear the burden of navigating unfriendly and uninformed waters (at home, school, and in the medical arena) to support their intellectually and developmentally impaired family member(s) (6,7).

Children with intellectual or behavioral disabilities have numerous challenges in their lives, including (but not limited to): late or absent development of appropriate language and communication skills; inability to navigate their environment without injury; impulsiveness with lack of self regulation and inability to discern cause and effect; anxiety, fear and limbic rages (sensory-sensitivity); and an inability to develop social skills.

Service dogs have long been utilized for individuals with a diversity of diagnoses and are now serving with ever-growing anecdotal evidence of effectiveness for additional conditions (8). Previously common services include; guide dogs for the blind or visually impaired, mobility assistance for individuals who are wheelchair bound and/or paralyzed and in need of having items retrieved, support for maintaining balance, hearing assistance dogs, seizure-alert dogs and service dogs for individuals with post traumatic stress disorders or other psychiatric illness (9).

This case report informs individuals affected by intellectual disabilities, their families, caregivers, health care providers and those focused on the study of and care for persons with intellectual or behavioral disabilities, of the availability and perceived effectiveness of service dogs as a treatment or therapy modality.

A story

This case report provides a brief introduction to the types of service animal organizations, the kind of assistance provided by the dogs, the rules governing the support of assistane dogs for the care and improvement of affected individuals, and evidence of success for multiple conditions, including those living with FASD. This case report specifically includes the informal observation of the recipient family to collect information on the impact of the addition of the specially trained canine (Chancer, a golden retriever).

Pre-intervention assessment was summed up in the comments made by adoptive parent, Donnie Winokur, to the CDC Regional Training Centers. Her son, Iyal was adopted in 1999 at the age of 14 months from Astrakhan, Russia. She stated, “In our experience, FAS is like the undertow…it lies in relative silence beneath the surface of the individual gathering its energy and force. Marked by co-existing disorders and peculiar characteristics, the ebb and flow of these behaviors are not exactly invisible, but can remain hidden. Some symptoms can be as gentle as a tide or swell like a tidal wave, engulfing the individual and often the caregivers or family. Living with a child affected by prenatal alcohol exposure is like living with the constant anticipation of an impending hurricane. Witnessing your child drown in a whirlpool of chemical imbalance and then resurface, keeps you hyper-vigilant for finding safe ports in the storm" (10).

Post-intervention assessment

"If nothing ever further changes as a result of Chancer (service dog), we have seen enough improvement that the effort and investment has already been worth it!" (Parents of Iyal). Since receiving their service dog, the family has noticed a decrease in Iyal's emotional outbreaks (frequency and severity); an increase in his ability to go to a restaurant and remain peaceably for 2+ hours; improved self-reflective thinking (opinion and judgment); impulsivity (decreased occurrence of darting into the street); increased "other" focusing. (theory of mind). In addition, social interaction has improved as the assistance dog serves as a social lubricant or icebreaker. The assistance dog helps with behavior disruption (a "nuzzle" command informs the dog to break through Iyal's own physical barrier and lick his face which comforts and calms him while deescalating the chemical storm), minimizing over arousal (providing deep pressure by lying (85#) over him, and being petted by Iyal); and a dramatic increase in expressive language using multi-syllabic words in complex sentences. Chancer "alerts" if Iyal disappears from his vision, and rarely lets him out of his sight. If he hears Iyal starting to cry or express frustration, Chancer gets up from wherever he is in the house and seeks Iyal out.

Of interest, the required training of dogs and child-parent pairs for certification and family placement occurs over months. The service dog agency from which Chancer was obtained (4 Paws for Ability) employs viewing videotaped examples of the types of emotional outbursts the child/ individual experiences, and other problems at home, school, and in the community. Trainers interpret the behaviors and create signals or commands based on the body language or sounds of the individual. The dog is then trained specifically to assist the paired individual. A "handler" must also be trained, particularly for minors and others incapable of self-care. Another example of assistance includes protection such as preventing the child from running into the street or wandering off (elopement). In this case, the dog's leash is held by the parent (handler) and the dog is tethered to the child with a special harness. By command, the dog lies down and will not move if Iyal were to try and run off, especially into the street or other hazard. This unique support serves to greatly calm the family by keeping the child safe.

Discussion

Service dogs can successfully impact outcomes for persons experiencing invisible "dis" abilities. Whereas one possibly may discern assistance need for a blind individual with a cane and dark glasses or wandering facial/eye movements, for individuals who are wheelchair bound, and for deaf persons who either have visible hearing aids or communicate through sign language; FASD affected individuals are often less visibly disabled. They may have high or normal IQ's (range 20-110), typical facial characteristics and body habitus; however they often demonstrate great irregularities in their developmental profile (11). With the training and placement of this particular service dog in 2008, the list of animal assistance categories continues to expand.

Military dogs are available through the Canines for Combat Vets Program (12). These dogs help veterans of wars, especially the increasing number of disabled veterans returning from Iraq and Afghanistan with partial or complete amputations, post-traumatic stress, and other visible and non-visible disabilities.

Hearing dogs are available to help alert individuals to crying babies, fire and other alarms, intrusions, doorbells, name calls, etc. These dogs are trained to make physical contact and lead their deaf or hard of hearing partners to the sound's source or to a safe space (13).

Therapy dogs (unlike assistance dogs trained for only one individual) travel to hospitals to assist clergy in their interactions with sick and dying persons and their families, as well as persons housed in nursing homes, hospices, extended care, and rehabilitation centers (14). These dogs help by providing unconditional love, assurance, and encouragement. They help lower blood pressure, decrease stress levels (adrenelin), and assist individuals who are wheelchair bound or otherwise differently abled.

Seizure Alert Dogs unexplainably recognize the scents or signs of upcoming seizure activity in their human. By being alerted, the individual can successfully preempt the seizure by taking appropriate meds, prevent injury by accessing a safe place to experience the seizure, or call for help (15). Seizure Alert dogs may sense changes in the chemistry and body of their human, and as a result can alert household members so the individual doesn't succumb to asphyxia or other injuries as a result of unattended seizure activity.

Balance dogs work with individuals needing help with stability, such as those with multiple sclerosis and Parkinson's disease. All assistance dogs by the nature of their training and support increase the confidence, social skills, and acceptance by others for whom they have been trained.

Any or all of these dogs serve to publicize the individuals' handicap or limitation in a positive manner. Through increasing the visibility, the complexity of the users' condition(s) is validated. There are numerous other disorders for which these assistance dogs are useful. Bipolar disorder, Asperger, anxiety, global developmental delay, Fredric's ataxia, Menieres, fibromyalgia, autism, seizure disorders (Dravet's Syndrome), Intellecutal Disability to name a few. It is necessary for the handler to be aware of the service dog statute in their given state, and to carry appropriate validation of the authenticity of the dog as a medical intervention. (16, 17). The cost of assistance dogs varies from $10,000 to $40,000 for obtaining a certified service dog. Many agencies offer support with fund raising. At 4 Paws for Ability, and some of the other identified organizations, all dogs receive obedience training through partnerships with correctional centers. This collaboration allows for healing of incarcerated persons and their provision of life-changing help to others.

Conclusion

At this time, most providers are not aware of this intervention for such a diverse group of conditions. Use of service dogs truly may assist educators, parents and therapists working with differently abled children and adults. The hidden disabilities common to FASD are neurological, behavioral, and emotional, but the availability of a specially trained companion supports the transitions, decreases the chaotic experiences, calms and settles emotionally unstable children, and increases their ability to handle responsibilities and interpersonal relationships. Self-esteem improves as children are more accepted by others with their special companion.

In the United States, there are approximately 120 service dog organizations at this time. As of 2004, there was only one state wherein

Medicaid paid for provision of service dogs for individuals, that being Montana. Some of the organizations help allay the costs of the training and placement of the assistance animals.

In June, 2013, the American Academy of Pediatrics endorsed the use of animal assisted therapy as an evidence-based intervention for individuals living with FASD. This same family case report was cited in their findings (18). This case report suggests the consideration of employing a service dog for improvement in the management and well-being of children and others with special healthcare needs. Review of available resources suggests that well planned, scientifically specific programs should be developed to truly evaluate what works and why in the use of service/assistance animals for care of the disabled.

Acknowledgements

This work was funded by the Centers for Disease Control and Prevention, Center for Birth Defects and Developmental Disorders (Southeast United States Fetal Alcohol Spectrum Disorders Regional Training Center, Atlanta, GA).

References

[1] Fetal Alcohol Spectrum Disorders Center for Excellence. Starting the conversation: Town Hall meetings on fetal alcohol spectrum disorders. Rockville, MD: Center Subst Abuse Prev, Subst Abuse Ment Health Serv Adm, 2004.

[2] Center for Disease Control and Prevention. Website: www.cdc.gov

[3] Southeast Regional Fetal Alcohol Syndrome Training Center. Atlanta, Ga. Accessed 2013 Jan 02. URL: www.sefasrtc.org

[4] Smith IE. FAE/FAS: Prevention, intervention and support services commentary on Burd and Jelson, Coles and O'Malley and Streissguth. In: Encyclopedia on early childhood development. Montreal: Centre Excellence Early Childhood Dev, 2003:1-8.

[5] Burd LJ. Commentary: Interventions in FASD: we must do better. Journal compilation. Child Care Health Dev 2006;33(4):398-400.

[6] Olson HC, Jirikowie T, Kartin D, Astley S. Responding to the challenge of early intervention for fetal alcohol spectrum disorders. Infants Young Children 2007;20(2):172-89.

[7] Fetal alcohol spectrum disorder. Part 2 Strategies and interventions fact sheet. MNASAP Minnesota Adoption Support and Preservation. Accessed 2013 Jan 02. URL: www.mnasap.org

[8] Moment by moment. Delta Society and Pet Partners and the use of service dogs. Accessed 2013 Jan 02. URL: www.deltasociety.org

[9] Moore C. Service dogs. Accessed 2013 Jan 02. URL: www.k9web.com/dog-faqs/service

[10] Presentation by Mrs. Donnie Kantor Winokur. CDC Regional Training Centers' meeting. Atlanta, GA: CDC, 2008 Aug 15.

[11] Bertrand J, Floyd RL, Weber MK, O'Connor M, Riley EP, Johnson KA, Cohen DE. Fetal alcohol syndrome: guidelines for referral and diagnosis. Atlanta (GA): CDC, 2004.

[12] Dogs for deaf and disabled Americans from the New England Assistance Dog Services (NEAD). Accessed 2013 Jan 02. URL: www.NEAD.org

[13] Assistance Dogs International, Inc. Website on guide dogs, hearing dogs and service dogs. Accessed 2013 Jan 02. URL: www.assistancedogsinternational.org

[14] Mariner's Lighthouse website on ministry dogs. Accessed 2013 Jan 02. URL: http://lighthouse.marinerschurch.org

[15] Mott M. Seizure-alert dogs save humans with early warnings. Accessed 2013 Jan 02. URL: NationalGeographicNews.com/News 2004 Feb 11.

[16] Laws affecting the use of assistance animals. Accessed 2013 Jan 02. URL: http://www. acb.org/arizona

[17] Georgia Code. Title 30. Handicapped persons. Rights of persons with disabilities. Accessed 2013 Jan 02. URL: http://web.lexis-nexis.com/research

[18] American Academy of Pediatrics. Accessed 2013 Jul 24. URL: http://www.aap.org/en-us/advocacy-and-policy/aap-health-initiatives/fetal-alcohol-spectrum-disorders-toolkit/Pages/Medication.aspx

Section 2: The School

In: Environmental Health
Editors: IL Rubin and J Merrick

ISBN: 978-1-62948-155-5

Chapter VII

Disaster preparedness in schools

Delon F P Brennen, MD*, MPH,
Myra Barnes-Eley, MPH and Michael P Poirier, MD
Department of Pediatrics, Morehouse School of Medicine, Atlanta, Georgia and Center for Pediatric Research, Children's Hospital of The King's Daughters, Eastern Virginia Medical School, Norfolk, Virginia, US

In times of heightened awareness regarding natural and man-made disaster threats, it is imperative that communities and families are prepared for catastrophic events. Research has shown schools are ill-prepared to manage many of these situations and parents are unaware of their children's schools disaster plans. This chapter was conducted to ascertain parents' current knowledge of school disaster preparedness. Methods: Investigator-administered surveys that addressed knowledge in areas of preparation, incident response, and recovery were given to parents of school-aged children who presented to a Pediatric Emergency Department (ED). Results: 340 parents were surveyed. 54% of parents felt that their children's school was not located at or near a potential disaster site. 60% knew of the existence of a disaster plan at the school and 80% did not know where their child would be in the event of an evacuation. 25% of parents did not know if emergency equipment was available at the school and 62% were not aware of the existence of a

* Correspondence: Delon FP Brennen, MD, MPH, 720 Westview Drive, SW, Department of Pediatrics, Atlanta, Georgia, 30331 United States. E-mail: delonbrennen@gmail.com

"crisis team". Over 36% of parents wanted to be involved in preparing or updating the schools' disaster plans. Conclusions: Even in the atmosphere of heightened disaster awareness, significant gaps exist in parental knowledge regarding their school's readiness for disaster. Improving parental awareness of the details of their school's disaster plan is necessary for adequate readiness for this very real threat.

Introduction

A disaster is defined as "an event that destroys property, causes injury or loss of life, and affects a large population or area" (1). As part of the Public Health Service, the Emergency Medical Services for Children's mission is to reduce "child and youth mortality and morbidity sustained as a result of severe illness or trauma" (2). At no time since the program's inception has that goal been more critical than the present.Current threats include not only conventional dangers such as fires, school shootings and weather-related disasters, but also biological, chemical, nuclear, and terrorist activities.

As of 2007, there are approximately 55.8 million students in the United States attending more than 119,000 public and private, elementary and secondary schools. Another 7.1 million adults work as teachers or staff in these settings (3). Based on these statistics, approximately one fifth of the entire United States population is in a school setting on any given weekday. These numbers help to underscore the importance of having focused and comprehensive crisis preparedness plans in place to protect citizens in these settings. However, in 2001, Sapien and Allen reported that school staffs have little training and equipment to deal with emergencies and, generally, are ill-prepared to care for students or staff that may need emergency care using specialized equipment and emergency training (4). Various other studies of school preparedness have given emphasis to medical emergencies and response, mainly limited to individuals but not groups of individuals or mass casualties (5-7).

The coordination and communication of designed plans to the parents of students attending these schools is an integral part of disaster preparedness activities. Parents play a consequential role in the success of any plan through their interaction with the students, volunteer activities in times of need, and cooperation with authorities and administrators in times of crisis. In an evaluation of preparedness of the 20 largest school districts in the United States, parents were included as part of the process, and their responses indicated that most parents continued to feel as though they did not get enough information regarding crisis management plans (8). Other studies of school preparedness activities failed to determine what parents knew of the school's plans or how involved parents were in preparing or executing the plans (9,10). Determining parents' knowledge base and attitudes towards disaster preparedness activities in communities, and in particular in schools, would provide a foundation on which to build and expand the partnership required for effective disaster planning.

Our study

This study used a cross-sectional design. Data collection occurred during a two-month period in the summers of 2004 and 2005 in the Emergency Department (ED) of an urban pediatric hospital. A member of the research team approached parents during their wait in the ED, as a convenience sample, and asked the parent if they had any school-aged children. If the parent had no school-aged children, they were not given the survey. If a parent stated that they did have a school-aged child, they were asked if they were willing to take a 10-15 minute survey about disaster preparedness in their child's school. The survey was anonymous and neither the parents nor their children could be identified later in the study by their responses. If they agreed, the survey was conducted in a private area, in an interview style, and the parents' responses were entered directly into a password-protected MS Access database. If the parent had more than one school-aged child, the parent was told to answer the questions in regards to their oldest child in school, because this would have allowed them more time in the school system and more potential exposure to school disaster plans.

Survey instrument

A 22-question survey was developed by the research team to assess parental knowledge in three areas of disaster preparedness activities at their children's school: 1) Pre-Incident Preparation, 2) Critical Incident Response, and 3)

Post-disaster recovery

The Pre-Incident Preparation section asked parents if they were aware of the emergency preparedness resources available at schools, such as a school nurse, the nearest hospital and emergency response agencies, injury prevention and emergency courses, and the existence of a disaster preparedness plan.

The Critical Incident Response section includes questions about the process that would take place in the event of an emergency, such as the tasks of crisis team members, evacuation procedures, the availability of medical equipment, and communication procedures to alert parents and the community.

The Post-Disaster Recovery section includes questions about the parents' awareness of recovery procedures, the availability of psycho-social personnel for the students (eg. student counselors), and critical incident stress debriefing. Most questions could have been answered with a "yes", "no", or "I don't know" response. No parent demographics were collected on this survey.

Statistical methods

The reliability of the survey was assessed using a test-retest method. A sample of 20 ED parents were given the survey through an interview and were asked to complete the survey again on paper approximately 30 minutes after the oral interview. Agreement between the responses of the oral and paper survey was assessed by computing, for each question, the percent of parents that gave the same response on the paper and the oral survey. If less than 60% (fewer than 12) parents' responses did not agree on the paper and oral survey for a particular question, that question was reviewed to determine if it needed to be changed. For those questions that had only 2 possible responses, a Kappa statistic was computed, where good agreement was declared at a Kappa of 0.80.

A qualitative assessment was done to test the construct validity of the survey. Another sample of 20 parents were read each question on the survey and asked if the question was clear and understandable, and if not, how could it be changed. The survey was revised based on the findings from the reliability and validity tests. The results reported here are those obtained after survey revisions.

Survey responses

Frequencies and relative frequencies of each question's responses were computed. The chi-square test was used to determine if associations between categorical variables were statistically significant, where significance was declared at an alpha of 0.05. Statistical analyses were performed using the SAS System for Windows, Version 8.2 (SAS Institute, Cary, NC). This study was approved by the Institutional Review Board (IRB) at Eastern Virginia Medical School.

Findings

The test-retest identified 6 survey questions that had less than 60% agreement on the paper and oral survey. These questions asked parents about the schools' school's crisis management planning team, post-disaster recovery plan, and effective means of communication that could be utilized by the school to inform parents of disaster preparedness plans. The text of those questions were unchanged, however, they were enhanced by adding definitions of specific terms that may have been unclear to parents. Only one question was changed as a result of the qualitative assessment.

Table 1. Emergency Department and Survey Demographics

ED Demographics **Fiscal Years 2004-2005**		**Survey Demographics** **June – July, 2004 and 2005**	
83,702 patients seen		▪ 340 parents interviewed	
▪ 16,168 school-aged children (5-19 years)			
○ 5-10 years	56%	○ Parents of elementary school students	61%
○ 11-14 years	26%	○ Parents of middle school students	24%
○ 15-19 years	18%	○ Parents of high school students	15%

Survey results

There were 371 parents who were asked to participate in the survey. A total of 340 parents completed the survey interview. Those that refused most often stated that a lack of time or their focus on their ill child as reason for not participating. Ninety-two percent of the parents interviewed had children in public school. The Emergency Department demographics and school grade levels of the children of interviewed parents are shown in table 1.

Pre-incident preparation

It was presumed that parents would know that schools were prepared for common medical emergencies and accidents. They were asked if a school nurse was present at their child's school and if medical equipment was available. Almost all the parents (97.1%) stated that their children's school employed a nurse and over 71.1% said that the school had medical equipment available.

With the aim of assessing parents' perceived threat of a disaster, they were asked if their child's school was near an area that may be susceptible to any type of man-made or natural disaster. Over half (54.3%) of the parents answered "No" to this question. It was found that 31.2% of parents did not know if their child's school had a disaster preparedness plan in place. Parents' perception of the schools' susceptibility did not significantly affect whether or not they knew if a school disaster plan was in place ($p=0.869$). Fifty percent of the parents said that they and their children had their own disaster plan at home. A significantly higher proportion of parents with a disaster plan at home definitively knew whether or not there was a plan at their child's school ($p=0.0002$). Of the 59.7% of parents that said there was a plan at their child's school, many did not know the details of the school's plan, such as the presence of a crisis team (56.7%) or if the school taught emergency courses, such as first aid or CPR, to the staff (42.9%).

Critical incident response

In the event of a school disaster, it is important to be able to notify parents, treat the injured, and identify children incapable of confirming their identity. Over 96% of parents stated that their child's school asked for emergency contact information, their child's primary care physician, and medical history information. Almost 23% of the parents said that their child was required to carry a school-issued identification badge. It is also important to assure that entry to the school is limited during a disaster. Over 84% of parents reported that there was restricted access to their child's school, meaning that the general public must go through some type of security stop or identification process before being allowed onto the school premises. It may be unsafe for students and staff to remain on school grounds during a disaster, thus requiring that the premises be evacuated. Eighty percent of the parents said they did not know where the child would be taken if they school had to be evacuated.

Post-disaster recovery

Most of the parents (61.2%) did not know if there was a post-disaster recovery plan at their child's school. Of the 19.4% of parents that said there was recovery plan in place, the majority of them knew details about the plan, such as the availability of psychosocial personnel (93.9%) and the inclusion of critical incident stress debriefing (63.4%).

Other considerations

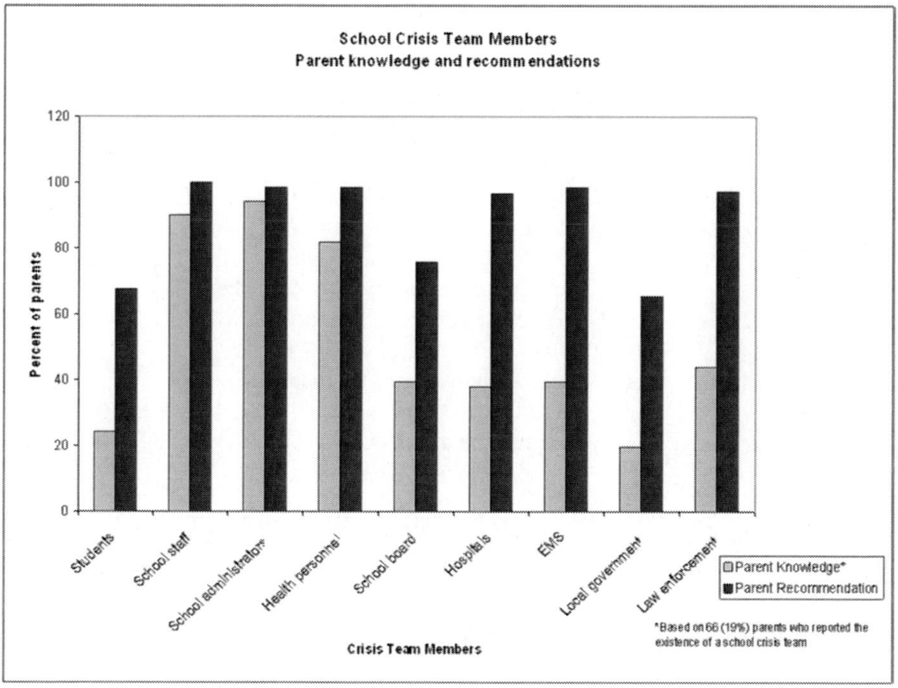

Figure 1. Parents' reported knowledge of current school crisis team members and their recommendations for crisis team members.

The parents had varied opinions about who should be involved in planning the schools' disaster preparedness activities. These results are shown in figure 1. Over 36% of the parents felt that they should be involved in preparing the schools' disaster preparedness plan. Parents were also asked how the schools currently communicated general information to them and how they would like to be informed of disaster preparedness information. Parents mostly received information by letters sent home from school (97.6%), but they reported that they would prefer varied methods, including face-to-face meetings (39%), television (11%), and email (10%). Parents suggested that disaster drills be performed at the school either monthly (49.7) or every grading period (40%). When asked how schools could better prepare for disasters, 43.5% of the parents suggested that parents should be kept more informed, 21.6% suggested having more frequent disaster drills at the schools, and 35.8% suggested that the children are educated more on disasters.

Table 2. Relationship Between Parent Demographics and Disaster Preparedness Knowledge

Demographic Characteristic	Associated knowledge	p-value
Higher Education level	Having a disaster plan at home	0.005
	Knowing where child would be taken during an evacuation	0.042
	Knowing about the existence of a disaster recovery plan	0.026
	Perceives that the school is located in an area at risk for man-made or natural disasters	0.012
	Preferring schools to communicate disaster preparedness information via the web[1] or email[2]	0.006[1], 0.002[2]
Two-parent household	Thinking that their child's school is in an area susceptible to a disaster	0.0006
	Having a disaster plan at home	0.033
Mid-income level	Thinking that their child's school is in an area susceptible to a disaster	0.0033

It was found that the grade that the child was in did not have a role in the knowledge that parents had about school disaster preparedness for most of the questions. The only significant observations made were that more parents of middle-school children reported not knowing if the school had a nurse (p=.0262) and if there was emergency equipment present at the school (p=0.0096). The parents' demographic data was collected during the second summer of the study. The results from that data are reported in table 2.

Discussion

Even before the most recent rash of terrorist events, The School Health Task Force of the American College of Emergency Physicians-Pediatric Emergency Medicine Committee identified mass disasters as potential emergencies that may affect those in school settings (11). While schools may or may not be direct targets of terrorism, they will certainly be affected by both natural and manmade disasters either directly or indirectly. For example, children may be injured by debris or left without a caretaker after a disaster. School settings may be also recruited to act as rendezvous locations, command centers, or shelters either during or in the aftermath of a disaster. Our results show that although the local society has been exposed to an ever increasing number of disaster-type events in recent years (hurricanes Floyd-1999, Isabel-2003 and Katrina-2005, school shootings in Littleton Colorado–1999, Beslan Russia - 2004, and Redlake Michigan-2005, school fire in Kumbakonam, India-2004, terrorism events in New York City and Washington DC – 2001) the perceived threat of a disaster happening in the local community is low, as the majority of parents in the study do not feel their child's school is at risk of a natural or man-made disaster. This information is even more startling when one adds that the community studied contains the world's largest naval installation, an air force base, industrial parks, large areas of coastal exposure, bridges and inland waterways. It is interesting to note that although there were some surveyed parents who had a higher perception of susceptibility; it did not translate into action, illustrated by the fact that higher perception did not produce a higher level of knowledge of school disaster plans. These results are consistent with other studies that showed perception of risk to be low not only in parents but also in school nurses charged with caring for children during these incidents both nationally and internationally (12,13).

Although it is assumed by many that community disaster preparedness activities will automatically involve or prepare school personnel for these events, it is interesting to note that schools are not mentioned once in the Department of Homeland Security's National Response Framework (14). This document is meant as a "guide to how the Nation conducts all-hazards response". This oversight underscores the need for schools to engage in active preparation activities, because they may not be included in community plans.

It is not a new idea that there is a need for schools to develop preparedness plans. McKenzie noted in 1980, "...if school personnel have taken the time to prepare written emergency care policies, they have at least recognized the importance of an emergency care program and the need for advanced preparation" (15). Before the recent increase in mass casualty and disaster events, school preparedness was mainly focused on issues of individual injury, along with playground and bus safety (16). In March of 2003, the US Department of Education suggested that schools individually develop a "crisis plan". This plan would be in partnership with "public safety agencies, including law enforcement and fire, health, mental health and local emergency preparedness agencies" (17). This collaboration would be fostered with the goal of ensuring that schools are prepared to protect the lives and health of students and staff in emergencies and disasters. The need for a plan is underscored by views that not all disasters can be handled by merely "calling 9-1-1". Helfand et al noted that without a plan "quick medical response [could] be hampered by mass evacuation, and local medical facilities, if not contaminated or destroyed, would become overwhelmed almost immediately" (18). To underscore the need for effective communication of these plans our results show that over a quarter of the parents did not know of the existence of a school disaster plan, while the majority of others did not know if a specific team was in place that was charged with responding to disasters that affected the school.

In January of 2004, groups concerned with the medical emergency response plans for schools such as the American Heart Association, American Academy of Pediatrics, American College of Emergency Physicians, National Association of School Nurses, and others, endorsed a statement paper aimed to "help schools prepare for life-threatening medical emergencies in the first minutes before arrival of emergency medical services personnel" (19). Statements such as these help to highlight the need for emergency plans in schools. The parents were generally familiar with the schools' knowledge and preparations for medical emergencies, as nearly all of them knew of schools asking for emergency contact, primary physician and medical history information. However, when asked about preparations for more disastrous events, the parents were generally "unaware". For example, an overwhelming number of the parents did not know where their child would be taken in the event of an evacuation, and many did not know what equipment was available at schools, or which community agencies were involved with the school's disaster planning activities and efforts. It has been noted that at a minimum "schools must be prepared to inform parents of the location of their children and how to reunite with their children" (20). Schools that wish to keep potentially sensitive information private, such as precise evacuation locations outside of public knowledge, for fear of increasing the risk to children in the event of terrorism, could inform parents that evacuation plans are in place, and for security reasons exact details will not be released until the appropriate time. This would potentially quell some parental anxiety about their children's safety. However, it is evident that many schhols have not addressed these issues as a National Survey of 3670 school districts noted that "almost one quarter have no written release plans" (21).

The major job of the school and school personnel after a disaster event is to return the school to its normal activities (17). The parental knowledge of post-disaster recovery plans to accomplish this goal was low, with more than half of the parents reporting no knowledge of a specific plan to address this need. Unfortunately the parental anxiety about how children will cope, and the possibility of maladaptive behaviors of children at varying stages of development, have the potential to be a volatile mix of emotions and actions. This mix could result in further damage to schools, homes, and communities. As David Alexander wrote, "This reaction is particularly tragic in response to disasters, in which incorrect beliefs often are the basis for misguided actions that lead to avoidable casualties and suffereing" (22). This could possibly be avoided with effective planning, and communication of those plans to the collaborating agencies involved.

In an attempt to discover whether a lack of knowledge was coupled with a lack of interest, we inquired about the parents' thoughts about disaster preparedness in schools, and possible recommendations for schools. We found that the surveyed parents have a high level of concern about these activities, and how their children will be cared for during these events. By open-ended questioning, some parents offered opinions and suggestions for improving the system: including more community agencies along with parents at the formative and subsequent stages of school disaster planning and on crisis teams, use varied forms of communication (cellular phones, radio, internet/email, face-to-face meetings) to get information to parents, have proper supplies available, listen to students and parents about violence issues in schools, and educate the children further on what disasters are likely to occur and what the schools preparations are for these events. Other studies have shown that parental involvement can be multi-factorial, from serving on advisory committees, helping to develop school safety plans, and donating equipment and supplies to fundraising for equipment and supplies, and supervising students during actual emergencies (23).

Limitations

A convenience sample of parents presenting to a pediatric emergency department was used. These parents presented to the ED for non-disaster related issues and non-emergent medical conditions, and so their responses to the interview should not have been influenced by their reason for being in the ED. The parental responses may have also been subject to a social desirability bias, as parents may want to answer the way they felt the interviewer wanted them to answer (saying they knew more about their children's' school than they actually did). An attempt was made to minimize this by informing parents that an "I don't know" answer was just as important as a "yes" or "no" answer. However, our results would then be an over-estimation of how much parents knew, rather than an under estimation, and thus continue to underscore the lack of knowledge within this group.

Since it was a convenience sample and drawn from one study center, the results may not be generalizable to all populations and communities. The issues raised by parents and the study may affect each community differently and each community may be more or less prepared for school disasters. However, each of the issues is important, and must individually be addressed in each and every community in order to maximize our schools' abilities to respond to disaster events.

Additionally, we had a small sample for the reliability and validity assessments. A larger sample may have yielded different results, thus altering more of the survey questions. Our study results may have also been different if conducted during the school year instead of during the summer. However, we considered this to be the ideal time to survey parents as they would have had the ability to have an entire school-year's worth of information on school disaster plans, and thus wouldn't have possibly been informed about these plans after the survey was completed, but before the end of the school year.

Conclusion

Our results show that the majority of parents with school-aged children may have little idea about the schools' disaster preparedness plans or what roles they and their children play. It is a natural instinct for parents to attempt to protect their children, and in many situations they will place themselves in danger to further this aim. In a disaster situation this could prove risky, if not fatal for the children, the parent, and other personnel charged with the duty of containing the crisis. Increasing outreach and educational training for these parents, as well as encouraging their input into disaster planning, may help with the goal of producing an informed and involved school population and community, and ultimately, maximizing survival and minimizing biological, physical, mental, and emotional morbidity that could result from a school disaster.

Future recommendations

Further research needs to be undertaken to ascertain whether the amount of knowledge parents possess about their child's school's preparation for disaster translates into different behaviors during a disaster event. This could be integral in using what parents know about the plans in order to produce the best possible outcomes during these tragic situations.

Acknowledgements

The Primary Care Infrastructure, Development, and Education (PRIDE) grant at Eastern Virginia Medical School, which is funded by the Health Resources and Services Administration (HRSA), provided services for epidemiology/biostatistical consultation and survey design. The Emergency Department at The Children's Hospital of The King's Daughters provided the patient population for this study. The Department of Pediatrics at Eastern Virginia Medical School provided faculty research development funds to support students in the Center for Pediatric Research Summer Scholars Program. Special recognition is given to Rianna Barnes, BS and Susan McMullen, BS, medical students at Eastern Virginia Medical School and scholars in the Center for Pediatric Research Summer Program for their diligent work on this project. Sources of Support: The Primary Care Infrastructure, Development, and Education (PRIDE) grant at Eastern Virginia Medical School, which is funded by the Health Resources and Services Administration (HRSA), provided services for epidemiology/biostatistical consultation and survey design.

References

[1] Health Resources and Services Administration, Maternal and Child Health Bureau. Five-year plan: Emergency medical services for children, 2001-2005. Washington, DC: Emerg Med Serv Children Natl Resource Center, US Dept Health Hum Serv, 2000.

[2] National Advisory Committee on Children and Terrorism. Recommendations to the Secretary of Health and Human Services. Washington, DC: US Dept Health Hum Serv, 2003.

[3] National Center for Education Statistics. US Department of Education, projections of education statistics to 2016; and unpublished projections and estimates; August 2007; Accessed 2013 May 15. URL: http://nces.ed.gov/programs/digest/d07/tables/dt07_001.asp?referrer=report

[4] Sapien RE, Allen A. Emergency preparation in schools: A snapshot of a rural state. Pediatric Emerg Care 2001;17(5):329-33.

[5] Allen K, Bail J, Helfer B. Preventing and managing childhood emergencies in schools. J Sch Nurs 1998;14(1):20-4.

[6] Hohenhaus SM. Pediatric emergency preparedness in schools: A report from the 2001 Southeastern Regional EMSC Annual Meeting. J Emerg Nurs 2001;27:353-6.

[7] Hazinski, MF, et al. Response to cardiac arrest and selected life-threatening medical emergencies: The medical emergency response plan for schools. A statement for healthcare providers, policymakers, school administrators, and community leaders. Pediatrics 2004;113(1):155-68.

[8] Phinney A. Preparedness in America's schools: A comprehensive look at terrorism preparedness in America's twenty largest school districts. New York: America Prepared Campaign. 2004. Accessed 2013 May 15. URL: http://www. workplaceviolence911. com/docs/20040916.pdf

[9] Kano M, Borque LB. Experiences with and preparedness for emergencies and disasters among public schools in California. NASSP Bull 2007;91(3):201-18.

[10] Olympia RP, Wan E, and Avner JR. The preparedness of schools to respond to emergencies in children: A national survey of school nurses. Pediatrics 2005;116(6):738-45.

[11] Abrunzo T, et al. The role of emergency physicians in the care of the child in school. Ann Emerg Med 2000;35(2):155-61.

[12] Evers S, Puzniak L. Bioterrorism knowledge and emergency preparedness among school nurses. J Sch Health 2005;75(6):232-7.

[13] Satapathy S, Walia A. Affected parents' and other stakeholders' perception of a fire disaster management in India: A situational analysis. Disaster Manage Response 2007;5(4):111-8.

[14] US Department of Homeland Security, Federal Emergency Management Agency: National Response Framework Washington DC, January 2008. Accessed 2013 May 15. URL: http://www.fema.gov/pdf/emergency/nrf/nrf-core.pdf

[15] McKenzie JF. Is your school prepared to handle Emergency Care? J Sch Health 1980;50(7):416-8.

[16] Miller TR, Spicer RS. How safe are our schools? Am J Public Health 1998;88:413-8.

[17] US Department of Education, Office of Safe and Drug Free Schools, Practical Information on Crisis Planning: A Guide for Schools and Communities Washington DC, 2003 (rev.2007). Accessed 2013 May 15. URL: http://www.ed.gov/admins/lead/safety/emergencyplan/crisisplanning.pdf

[18] Helfand I, et al. Nuclear terrorism. BMJ 2002;324(7333):356-9.

[19] Hazinski MF, et al. Response to cardiac arrest and selected life-threatening medical emergencies: The medical emergency response plan for schools. Ann Emerg Med 2004;43:83-99.

[20] Graham J, Liggin R, et al. Planning for a mass casualty incident in Arkansas schools. J Sch Health 2005;75(8):327-8.

[21] Graham J, et al. Mass-casualty events at schools: A national preparedness survey. Pediatrics 2006;117:8-15.

[22] Alexander DE. Misconception as a barrier to teaching about disasters. Prehosp Disaster Med 2007;22(2):95-103.

[23] Nano M, et al. Are schools prepared for emergencies? A baseline assessment of emergency preparedness at school sites in three Los Angeles County school districts. Educ Urban Soc 2007;39:399-422.

In: Environmental Health
Editors: IL Rubin and J Merrick

ISBN: 978-1-62948-155-5

Chapter VIII

Health care service access and delivery to youth with special health care needs attending exceptional student education school

Farheen Akbar, BS, MPH, Jeffrey Alan Aenlle, JD, MPH, Katryne Lukens-Bull, MPH, Allyson Hall, PhD and David Wood, MD, MPH*

College of Public Health and Health Professions, University of Florida, Center for Health Equity and Quality Research and Department of Pediatrics, College of Medicine-Jacksonville, University of Florida, Jacksonville, Florida, US

Adolescents with special health care and educational needs face significant obstacles as they age out of traditional child health care and education. Transition services is the term used to describe a set of services and supports designed to assist adolescents in adjusting to the change from school to work, home to independence, and pediatric to

* Correspondence: Professor David Wood, MD, MPH, UF College of Medicine-Jacksonville, Jacksonville Health and Transition Services (JaxHATS) and Center for Health Equity and Quality Research, 580 W 8th St, Room 7005, Mailstop T-60, Jacksonville, FL 32209 United States. E-mail: david.wood.

adult health care. This project partnered with three Exceptional Student Education Centers (ESE) as part of the Duval County Public School system to study the current health and social needs and health care transition needs of adolescents ages 16 to 22 years with complex developmental and health care needs. The study included a survey of parents of adolescents with special health care needs, in-depth key informant interviews and focus groups. A majority reported that their youth has a doctor or nurse who knows them well, a quarter indicated mental and dental services were needed. With respect to transition services, 60% of parents said that their adolescent's health care provider discussed how health care needs would change as the youth approached adulthood. However, only 1/3 of parents had received information on obtaining health insurance or reported that a provider developed a plan to address changing needs. In-depth interviews and focus groups revealed barriers to care which included actual physical needs of the adolescents, cost of medications, difficult social environments and lack of coordination of services. Ideas for reforms of the system included campus health clinics and psychiatric consultation, and health care transition guides for parents.

Introduction

The Maternal and Child Health Bureau of the Health Resources and Services Administration in their 2005/06 report found that of the approximately 27.1 million young adults aged 18 to 24 in the United States, 16.8% had a special health care need (approximately 4.6 million adolescents). Children with special health care needs are "those who have or are at increased risk for a chronic physical, developmental, behavioral, or emotional condition and who also require health and related services of a type or amount beyond that required by children generally" (1). These adolescents face many barriers as they transition to adulthood. When these adolescents have serious, complex health issues, reside in segregated, vulnerable communities fraught with disadvantage and poverty, or are cared for by family members or caregivers who are not able to effectively advocate for their health and independence as they become adults, they often suffer greater health and economic consequences.

Perhaps one of the most significant barriers to successful transition to adulthood faced among these youth is the fact that many of them live in poverty. Approximately 14% of all children with special health care needs live at or below 100% of poverty (2). Children with physical and mental special health care needs are more likely to live in poor families compared to children in general (3) and the chance of being at or below poverty increases for youth and young adults in this environment (4). Poverty has been shown to limit their access to needed specialty care, preventive dental health care, mental health services, therapy services, prescription medications, transition services, and access to care in general (5). Poverty coupled with low parental education has been shown to have a pronounced negative effect on access to and use of specialty care services (3).

Studies have reported that adolescents with special health care needs face a complex and challenging period that has recently been termed "emerging adulthood" (6). During this "emerging adulthood" stage of development (approximately ages 16 to 24 years) individuals typically will leave high school, pursue some combination of additional education, training or work, begin to establish long term relationships and leave their family home (6). For adolescents who have chronic health conditions or disabilities, this period of transition can be especially challenging and can negatively impact their health and health care (7). When these adolescents with special health care needs have the additional burden of residing in vulnerable community this transition can be even more difficult. Further, these adolescents must also navigate transitions from a familiar world of pediatric providers to a very different system of adult primary and specialty care (8). The combination of a youth's developmental immaturity, intense life changes, social environment and health system changes poses many potential threats to these emerging adults' (or their parent's) stable management of their health.

The health care system supporting the transition of adolescents with chronic health conditions from pediatric to adult care and other adult roles is complex with many potential barriers to successful transition. First, most primary and subspecialty pediatric providers stop serving children once they reach the age of 18 or 21 years and only a minority report creating plans for youth to access appropriate adult provider (9). Second adult health care providers are inexperienced in caring for persons with developmental disabilities (e.g. cerebral palsy, spina bifida, cognitive impairment) and/or complex, multisystem conditions arising in childhood (10). Third, the eligibility rules for various programs change as individuals get older, complicated reapplication processes are required, and the adult service system lacks various entitlements. Most forms of public health insurance for dependent children terminate at age 18 years (Medicaid), 19 years (Title XXI), or 21 years (Title V special funds) leaving the youth with no source of payment whatsoever for needed health services (11). As a result, youth are dropped from service and may not have adequate health insurance to support their needs. Finally, eligibility for services that are available varies by agency. Support staff and families may not be aware of the services for which youth are eligible.

There is a dearth of programs for youth with complex medical or developmental disabilities and there is currently little coordination between health care systems and the educational systems that serves these youth. As a result many youth even prior to transition do not receive adequate health care and experience health care complications. According to the National Survey of Children with Special Health Care Needs (NS-CSHCN 2005/06), 16% of children with special health care needs reported not receiving at least one service they needed in the past year. The most commonly reported service not obtained was preventative dental care. Children from low-income families or who are uninsured are at even greater risk for not receiving necessary care and are therefore likely to enter adulthood at an increased disadvantage. The NS-CSHN reports that 25% of children from low income families and 45% of uninsured children did not obtain at least one health service that was needed in the last year. For children with public insurance 22% did not receive at least one service in the last year compared to 11% of children with private insurance (2).

Among individuals with special health care needs, a particularly vulnerable group is youth with cognitive impairment, cerebral palsy, spina bifida and other conditions, who comprise 2-3% of this population. Aside from their chronic health care conditions, individuals with developmental disabilities are more likely to live in families that face a number of socioeconomic pressures. They are also more likely to be members of single female headed households and of families with lower incomes (12). In addition, mothers of adolescents with developmental disabilities have reported that working outside the home is often in direct conflict with their care giving responsibilities resulting in reduced employment and wages among these women (13).

One of the most significant trends affecting this group is the deinstitutionalization or movement of care away from large state institutions to community-based settings. In the United States, between 1980 and 2005, the daily census of individuals with developmental disabilities living in large institutions declined by about 69% (14). While this trend is positive from the standpoint of Healthy People 2010 goals for individuals with disabilities, it also exposes the fact that the social and health care safety nets do not have the capacity or resources to provide adequate support for these individuals and their families. Perhaps the greatest examples of this are the waitlists maintained by state Medicaid agencies for enrollment into Home and Community Based Services (HCBS) waiver programs. These programs are designed to prevent institutionalization by providing social support and home and community based services to individuals with disabilities (15). However, fiscal concerns often force states to limit enrollment and to use waitlists when the number of applicants exceeds the number of available slots. In 2005, waivers for individuals with mental retardation or developmental disabilities had approximately 138,000 people on their waiting lists, a 48 percent increase from 2004 (16). Being on a waitlist indicates that adolescents and their families are not receiving key supportive services.

Since adolescents living with a disability, especially those with cognitive or developmental limitation have a greater likelihood of living in poverty and without supportive community services; their transition from childhood to adulthood may be especially difficult. This has significant consequence for their health and social condition. Understanding the needs of low-income transitioning youth, their parents, and the educational system's services for these youth is important to developing policy and programmatic interventions to improve their health and educational outcomes. The purpose of this chapter is to describe a study of the health care needs and the transition services that are currently provided to youth between the ages of 16 and 22 years who are living with complex developmental and medical problems.

Our study

Duval County, Florida was chosen as the setting for this exploratory study of youth with special health care needs because it typifies the problematic interactions of poverty and transition to adulthood. Duval County is a large geographic area encompassing over 773 square miles with a population density of 1,006 people per square mile (851,000 total population). It has been synonymous with the city of Jacksonville, Florida since 1968 when the city and county governments were consolidated. Sixty four percent of the county is reportedly White while 29% is reportedly Black, with an additional 6% of people reporting other racial categories (17). Unlike other areas in Florida, Duval County has a fairly low Hispanic/Latino population overall. Although most of the county enjoys relative comfort with the median household income reported at $40,703, the county is economically and racially segregated into areas of high and low income, high and low opportunity, and high and low functional status and health (17). In March 2008 the Duval County Health Department issued the report entitled "Place Matters" to describe the health planning and geographical analysis to identify priorities areas for health promotion. The report delineated the county into six health zones (see figure 1).

These zones have allowed the health department to determine which areas have a greater health burden and may have other issues like poverty. Zone 1 comprises Duval County's urban core and has more minorities, people with lower socioeconomic status, individuals with lower educational attainment and lower health outcomes. Furthermore, census data have shown that those living in HZ1 have the lowest average household income at $21,875, with a poverty level of 28% while those living in HZ3 enjoy the highest average household income at $53,972 with the lowest poverty level at 5.3%. Further, 41% of all Blacks in Duval County live in HZ1 with approximately 80% of HZ1 reporting being Black (18). The implication of this segregation within the urban core community is that it creates a climate of community vulnerability and increased stress (18).

	Health Zone 1	**Health Zone 2**	**Health Zone 3**	**Health Zone 4**	**Health Zone 5**	**Health Zone 6**	**Duval County**
Poverty	28%	8.8%	5.3%	11.7%	10.8%	7.3%	11.9%
Education level high school or greater	63.7%	87.2%	92.5%	82.6%	75.7%	89.6%	82.9%
Children in poverty less than 18 y/o	38.4%	12%	6.4%	16.6%	14.5%	9.2%	16.4%
Average median household income	21,185	44,509	53,972	39,610	42,040	44,765	41,118
African American population	79.2%	19.8%	9.3%	21.4%	27.7%	10.8%	27.8%

Figure 1. Comparison African American area of city (HZ1) to other Areas.

Within HZ1 resides two of the three Exceptional Student Education Centers (ESE) of the Duval County Public School system. The Duval County School System serves over 17,000 students with special health care needs which is approximately 14% of the approximately 123,200 students who attend public school in the county (19). Of these students, approximately 500 attend one of the three ESE Centers (Duval County Public Schools). Students in these programs have special health care needs including conditions like moderate to profound cognitive impairment, Spina Bifida, Cerebral Palsy, seizure disorders, Muscular Dystrophies, Down syndrome, other genetic or chromosomal disorders, autism and other developmental disabilities. Because of their significant health issues, these children are not mainstreamed into the other public schools in the School District. Further, because of their significant educational issues the students may attend until their 22nd birthday.

The ESE Centers disproportionately serve families from the urban core of Duval County. Most adolescents at the ESE Centers and their families reflect the same socioeconomic characteristics of Health Zone 1. They are disproportionally poor; a majority receives free or reduced lunch, and most are on Medicaid. The racial/ethnic composition is approximately 60% African American and 40% Caucasian (19). School staff reported that they believe 99% of adolescents who attend their school are below the poverty level, the exception being one school outside of Health Zone 1 which reported approximately 50% of their adolescents living below the poverty level.

ESE Centers have full-time nurses, occupational therapists, physical therapists, speech therapists as well as other allied health professionals on site in order to provide care for the children and youth at the school. One location serves mostly young adults who have complex physical health needs. Many are not ambulatory and are in wheelchairs. However, many are medically fragile, and have complex medical and nutritional issues. A second location serves mostly young adults with behavior and mental health conditions such as autism or intellectual/cognitive disabilities. Students at the third location have a combination of both physical and behavioral and mental health issues.

This descriptive study consisted of a 42 item survey questionnaire to solicit information from families of students' ages 16 to 22 years old. In addition, in-depth interviews and focus groups with key informants in individual and group settings at each of the three schools and at the District administrative offices were conducted.

ESE Center staff worked with the study team to develop the survey instrument. The survey questions were also based on those questions from the National Survey of Children with Special Health Care Needs 2005/06 (2). The survey asked about current health care needs, whether those needs are being met, and what services students may need in the future as they transition out of pediatric care and the school system to adult health care providers. Four components of transition that were specifically assessed included: whether doctors had discussed the shift to adult providers; whether doctors had discussed the youth's changing needs as he or she approached adulthood; whether anyone had discussed health insurance coverage in adulthood; and whether the youth was encouraged to take responsibility for his or her health.

The survey was sent home to parents of 214 young adults attending the three ESE Centers. Parents were asked to complete the survey and return it to their youth's teacher. Reminder notes were sent home by the teachers and at two schools, parents were called and asked to complete the survey over the phone with the teachers. Additional surveys were available on-line and at the school. Survey responses were anonymous. Parents were offered $5.00 gift cards to a local grocery store for returning their survey and teachers were provided $25.00 gift cards for assisting with the study. Principals were also given $25.00 gift cards for their assistance with the survey and the qualitative data collection. Data was entered into an ACCESS database and then exported to SPSS 17.0. Univariate and bivariate analysis was completed using SPSS. Descriptive statistics were compiled and utilized with qualitative data and are presented in the results section.

In addition to surveys, a series of focused group interviews were conducted with each of the three Duval County ESE program staff. These interviews included members of the Individualized Education Plan (IEP) planning team, principals, school nurses, psychologists, teachers and guidance counselors. At the district level, a focus group interview was conducted with the Duval County ESE director and the behavioral health program director and staff.

Additional key informant interviews were conducted individually with the medical director and the nursing director. These interviews collected information on the transition planning process, the level of health care related transition support that the young adults receive, and the difficulties in supplying the student with key services. In addition, participants were asked to suggest ideas for reforming the system.

Three members of the research team served as the primary note takes during all interviews. The notes were reviewed and details added by the three note takers during discussion sessions after the interviews. Notes were then transcribed. The qualitative analysis relied upon thematic coding to summarize ideas that were implicitly and explicitly documented within the combined (from three note takers) transcriptions from the interviews conducted. Information was color coded within the text in Word in different colors or multiple colors using the highlight function to identify common ideas and concepts. For example, when lack of dental care was mentioned, this would be highlighted in green throughout the notes to indicate an issue with dental care but would also receive an additional color of pink to show an access to care issue. These colors were then grouped together to get an overall summary of the ideas across interviews. Recursive abstraction was then used to create three categories with subsumed themes within the first of the three categories (20). The three categories and their sub-themes include: 1) the barriers to the young adults' receipt of health services with sub-themes of a) the young adults physical needs, b) the young adult's social environment, c) barriers within the school system, d) barriers due to the health care provider; 2) the disconnected relationship amongst stakeholders and 3) ideas for reform.

Findings

A total of 214 surveys were distributed to the parents at the three locations. Forty eight surveys were returned representing a response rate of 26%. The two schools where teachers were able to call parents and do the survey over the phone had higher response rates (39%) than the school in which teachers were unable to do this. Principals and teachers thought that the low response rate was due to the fact that many of the parents may have had difficulty completing due to low literacy. Principals and teachers reported that they have had similar difficulties obtaining information from parents of children in the school and were therefore not surprised by the low response rate.

Over two-thirds of the young adults whose parents completed the survey were over 18 years of age, over 50% were male, and 50% were Black (see table 1). The majority of surveys were completed by parents, with only a few completed by grandparents. Almost half of the parents had greater than a high-school education (46%). Parents reported that about one-third of the young adults had health problems that affected their ability to attend school. Fifteen percent of young adults reported missing more than 20 days from school in the past year due to their health condition.

Access to care

Almost all of the respondents (94%) said that their youth has a doctor or nurse who knows them well and who is familiar with their health history (that is a medical home) (see table 2). Close to 90% of parents reported that their youth visited their primary care doctor at least once in the past 12 months. Parents were asked if their adolescent had needed health services but had forgone or delayed them in the past year. Twenty-five percent reported that their youth needed but did not receive mental health care and services (psychologist, psychiatrist, behavioral services and other mental health provider), and 23%

reported needing but not receiving dental care. When asked why care was not received parents reported difficulties associated with finding a clinician who would provide services for their youth's health problems (29.2%), lack of health insurance coverage (25%), difficulty finding a provider who would take Medicaid (22.9%), and not having money to pay for services (20.8%). Long waiting period was reported to be a barrier for about 13% of the respondents.

Table 1. Demographic Characteristics of Adolescents and Parents

(n=48)	**Percent**
Student Age	
Less than 18 years old	35.40%
18 years or older	64.60%
Gender	
Male	57.40%
Female	42.60%
Race	
Black	56.30%
White	33.30%
Other	10.40%
Person completing survey's relationship to student	
Parent/Guardian	93.80%
Grandparent	6.30%
Education level of parent/guardian	
Less than High School	8.30%
Completed High School	45.80%
More than High School	45.80%
Health Affects Ability to Attend School	
Some or a great deal	35.4%
Very little or not at all	64.6%
Days Missed Due to Health Problems	
None	17.0%
Missed 1-9 days	38.3%
Missed 10-19 days	29.8%
Missed 20 days or more	14.9%

Although the survey was close-ended, parents often "wrote in" or provided additional anecdotal information. The following quotes indicate the frustration and problems parents have experienced:

> "Medicaid insurance/ dental services are deplorable."
> "Some physicians don't want to see handicapped."
> "Dental service that is able to do more than cleaning. My child needs to be put to sleep in order to have dental work done."

Table 2. Access to care

(n=48)	**Percent**
Youth has a personal doctor or nurse who knows them well and is familiar with his/her health history. (Medical Home)	93.8%
During the past 12 months at least one visit to a primary care physician. (Pediatrician, Family Practice, Internist)	87.5%
Youth delayed or had forgone care for the following needed service.	
Mental Health Services	25%
Dental Care	22.9%
Gynecology	8.3%
Cardiology	6.3%
Parents were asked why they had delayed or forgone care	
Finding a provider who would care for their youth	29.2%
Lack of coverage for the service by my youth/young adult's health insurance	25%
Finding a provider who will take Medicaid	22.9%
Did not have the money to pay for the service	20.8%
Long waiting period	12.7%

Health insurance coverage and other benefits

Health insurance is one of the strongest indicators of access to health care (21). The majority of the parents (91.5%) surveyed reported they had some form of health insurance coverage for their adolescent: 39.1% had private insurance, 50% were on the Medicaid waiver program, and only 8.5% of parents reported that their adolescent was uninsured (see table 3).

Table 3. Health insurance coverage

(n=48) Insurance Type	Percent
Dental Insurance	62.50%
Medicaid Waiver Program	50.00%
On Waiting list for Medicaid Waiver	22.90%
Private health insurance	39.10%
Medical Services (CMS)/KIDCARE	17.00%
Military insurance (e.g. Tricare)	6.40%
Uninsured	8.50%
Ever had Social Security Income (SSI)	70.20%
Lost SSI at some point	33.30%

Further, 62.5% of parents reported that their adolescent had dental insurance. As in other studies (21,22) this study found that reporting having insurance did not ensure that the insurance was adequate to cover the health care, including dental care, which 23% of parents reported was needed by their adolescent. Further, all adolescents who attend an ESE Center are eligible for the Medicaid Waiver Program which provides them with a care coordinator to assist them in achieving their goals and provide resources for the coordination of their care, including health and dental services. Fifty percent of adolescents were reported to be receiving the waiver's services but an additional 23% were on the waiting list (which often can last over 7 years) and 27% had not yet applied for services through the waiver. Large portions (70.2%) of these adolescents were receiving Supplemental Security Income (SSI) but 33% of them had lost their SSI benefits at some point in their lives.

Transition services

Lack of transition support/services during adolescence has been shown to negatively impact the general health and future potential of adolescents with special health care needs (7). Sixty percent of parents responded that their adolescent's health care provider had discussed how health care needs would change as the adolescent approached adulthood (see figure 2). However, only 35% indicated that someone had discussed how to obtain health insurance as their young adult aged and only about 30% reported a provider developing a plan to address their young adult's changing needs. Most parents coordinated their own youth's health care and over half of them reported that they would like extra help in these coordinating efforts. About 40% of parents also noted that health care providers need to improve communication with the school system in order to better coordinate their youth's health.

Group and key informant interviews

Qualitative interview results and analysis revealed three themes and sub-themes 1) the barriers to the young adults' receipt of health services with sub-themes of a) the young adults physical needs, b) the young adult's social environment, c) barriers within the school system, d) barriers due to the health care provider; 2) the disconnected relationship amongst stakeholders; and 3) ideas for reform.

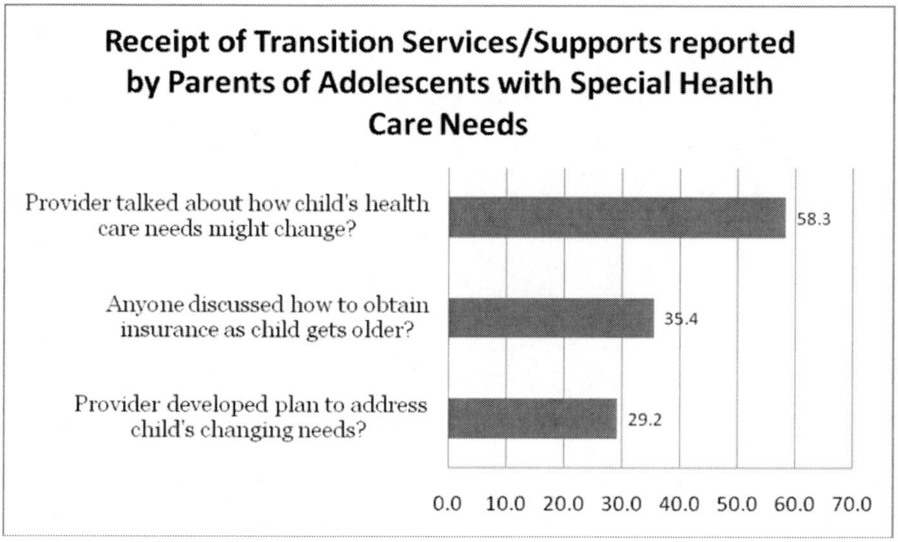

Figure 2. Transition services.

Barriers to the receipt of services: Physical needs

Principals and their staff (teachers, nurses, occupational therapists, etc) described difficulties associated with the physical needs of the students. The young adults' physical needs were mentioned as a barrier to care. For many of the young adults, their ability to participate in their own treatment is limited and many cannot complete basic self-care activities. One of the needs heard often from the school staff was how parents struggle to supply adult diapers for their adolescent. The schools cannot supply this and most parents cannot afford the cost (approximately $.83 per diaper if ordered in bulk). Parents also have a difficult time with adolescents who are mobility impaired. The schools have specialized equipment like Hoyer lifts for the larger adolescents whereas parents do not have these resources at home and must rely on neighbors or family to assist with basic care. In this, as in many areas, the schools were looked to as the health care provider.

School staff discussed parents' experiences and frustrations in trying to provide medications for their young adults. They stated that parents often are not able to pay for the medications and consequently, they may not give the schools the full recommended dose to administer to the adolescent. Teachers reported being told that parents had no medications at all or that they should "share" medication between siblings or friends. One teacher said, "Parents may give less med(icines) because they do not have the money for enough of them that month." Another teacher said that she knew of a case where a stepmother regularly sold the student's medication on the street. Unfortunately, this is not an uncommon problem where basic needs must be met and medications are a low priority.

Impact of social environment on youth with disabilities

Beyond their physical needs, respondents noted that these adolescents also have to learn to cope with issues that concern all teenagers such as sex, drugs, increasing responsibility, and decreasing parental and school support. These issues are impacted by social environment the youth lives in – primarily youth in the urban core live in economically disadvantaged neighborhoods. In many cases, the adolescent's ability to comprehend the dangers and consequences of their choices may be limited. This can place the adolescent and their classmates at risk. School staff provided multiple examples of instances where the young adults were at risk or placed others at risk due to "normal" adolescent behaviors that were compounded or exacerbated by the interaction between the adolescent's special health care need and social environment. Examples included a student who tried to sell bags of flour on the street to make money, imitating other teenagers within their own community selling cocaine. The school administrator offered that," "…the boy is gonna get himself killed."

School officials also expressed concerns about issues around sexual promiscuity on and off campus. School officials stated that approximately 10 pregnancies had occurred over the past three years in one school. Although lacking in the resources to deal with these issues, the school allocated staff to provide one-to-one supervision for some students to prevent inappropriate sexual behavior from occurring. School officials were concerned with issues that affect their adolescent students, especially after graduation when they would be "on their own" in the community. One school administrator stated that she had to testify in court on behalf of a young adult who, after graduation, was prosecuted for sexual offenses within the community Teachers have little leeway in addressing these issues at school because they are not allowed to counsel the students on specific aspects of sexuality.

Mental health issues and the social environment also interact, along with lack of access to health care, to create situations where adolescents are put at risk. Schools officials stated that multiple adolescents have been 'Baker Acted' (taken by police to a mental health facility) during school hours. One event was described as the police standing in the front office while the social worker held the door shut to a student who was exhibiting threatening behavior. Police informed the school that the mental health facility would not admit this student again because the student had been there many times before. The school had no resources for psychiatric assistance and neither the physician nor the parent responded to telephone calls. The only recourse for the school was to attempt to isolate the adolescent until they could find support from the parent or a medical provider. Examples of lack of access to mental health services were repeated over and over again at all three schools.

Barriers to the receipt of services: Within the school system

Each school spoke about their inability to maintain active parent-teacher associations due to lack of participation among parents. Most parents, according to school staff, were just overwhelmed from working and dealing with their adolescent's health care needs over many years. The school staff indicated that by the time the student was an adolescent, the parent was used to the system. They trusted the school staff, many of whom had educated and cared for their youth for at least five years prior to their graduation.

The inability to provide or coordinate mental health care services was also discussed among school leaders. For instance, there are no psychiatric services available for young adults who have cognitive disabilities and need mental health services to be able to participate in school (non-cognitively disabled adolescents are provided mental health services through the school system). These young adults must have their medications and behaviors managed outside of the school system, where there is a lack of providers.

There are also legal limitations on the amount of care that these schools can provide. Schools are not equipped to provide extensive medical and therapeutic care to young adults or to act as the primary or mental health care providers for adolescents with special health care needs. In terms of care coordination, schools must interpret physician requests based on what is necessary for these adolescents to participate in class. Parents become frustrated with the lack of resources and services in the schools, but the school's primary mission is not to provide health care but to educate.

Beyond health care, there are other limits to the resources schools provide. Vocational rehabilitation can assist with job placement through a guidance counselor employed at the ESE Centers. However, this can only occur within 90 days after graduation. Alumni are therefore faced with job loss and the lack of skills to ensure long-term employment success.

Barriers to the receipt of services: Due to lack of health care providers

Barriers also exist within the health care system. In Duval County there is a lack of health and mental health care providers who are also willing to take Medicaid and who possess the skills needed to care for adolescent with special health care needs. The school staff provided multiple examples of parent's inability to find mental health care or dental care within the county for their adolescent. Adolescents who need to be sedated or who have behavioral issues currently must travel to Gainesville, which is approximately 90 miles away, in order to see a provider at the University of Florida that will accept Medicaid. Some teachers stated that a student arrived at school with a swollen face due to an abscessed tooth. The nurse reported that the student was not eating and was obviously losing weight. School officials found that the parent could not afford to take the youth to Gainesville for dental care. The family had been to the emergency room. However, the emergency room could only stabilize the student and provide a referral. The adolescent was unable to actively participate in school and the nurse also stated she was afraid to send the student home on the bus due to her swollen mouth and its effect on her health.

Staff and school nurses often have to care for adolescents during the day who are ill (and did not have a parent come to pick them up). At the end of the day, the nursing staff had to make a medical decision whether the student could physically make it home on the bus without risking a medical emergency. Nurses stated that they had taken the bus home with students because of personal commitments to the health and well-being of the adolescents. Repeatedly, school staff spoke about how physicians in the community would not respond to their requests for medical intervention or for information. Often a student would be in immediate crisis and the school could not wait for 72 hours until their personal physician called them back. In many of these situations, the Duval County medical director, a doctor who is well known within the community, would also attempt to contact the student's physician. Despite his intervention physicians would often not return calls. Psychiatric services within the community are also inadequate to meet the need, especially for Medicaid recipients and those who have cognitive disabilities. One school administrator stated they literally have "nowhere to turn" for their adolescents who are having behavior issues. When an adolescent is being seen by a psychiatrist, the psychiatrists are even less likely than primary care physicians to return phone calls during crisis situations.

Finally, as some adolescent become adults, they move to group home settings, independent living settings, and other living arrangements other than with their parents. Often, staff or coordinators from these settings are unable, due to their company's policies or legal ramifications, to provide consent for treatments. One solution is to have an individual become a legal guardian for that adult. However, becoming a legal guardian is an expensive and complex endeavor and few take this option. As a result most young adults with cognitive impairment have no one to consent for their treatment. The health care system currently deals with these issues on a case by case basis, often resulting in delayed or forgone care.

Disconnected relationship among stakeholders

Study participants noted that there is a lack of communication and coordination across stakeholders involved in the care of the adolescent. There were reports that the parents did not truly understand the adolescent's diagnosis because health care providers do not communicate clearly with parents. One school teacher told us that

> "the health care system should tell the parents the truth about the child's problem. Parents think their autistic child is just going to wake up one day. They ask us why we are not doing more to wake their child up. Why are they not reading? I try to tell them that their child is not reading because he is never going to be able to read. We are trying to teach him to go to the bathroom on his own."

There were reports that some of the parents were themselves former ESE students and are unable to communicate with the school's faculty in an effective manner. One administrator told us that something that would "take 15 minutes to communicate to one of us, might take her 6 hours to explain to a low functioning parent." Further, the main method of communication between the parents and the schools is a note pinned on the backpack. This note is susceptible to being lost before even reaching the parent and requires the parent to have some level of literacy.

Ideas for reform of the system

Stakeholders were asked to think of creative ways to address the barriers they described. The ideas they mentioned centered around bringing health care services to the school, including on campus primary care clinics, on-site consultation by psychiatrists, and bringing the Duval County dental van to the schools. They also suggested that parents have access to more information on transition, such as a transition guide that would provide information on health care and other services for youth and young adults. Another suggestion was to have the primary health care provider appear at the adolescent's IEP meeting so that an open dialogue could occur, health needs would be accurately represented, and clear and concise orders from the physician could be articulated. The most optimal solution, from the key informants' perspectives, would be the full integration of physical, mental and dental services at ESE Centers with health care provider participation built into the IEP process allowing for fluid communication across all of the stakeholders.

Discussion

This study provides a description of the challenges faced by youth and young adults with complex medical and developmental conditions and by the families and school personnel caring for them. By virtue of their existing health problems these youth are at significant risk for adverse medical and social outcomes. However, this risk is exacerbated because they live in poverty and their social, educational and medical safety nets lack appropriate resources, are uncoordinated, and to some degree are simply unable to provide even basic services, such as restorative dental care. Our study illustrates gaps in the medical and mental health care system such as a lack of health care providers, lack of mental health providers, insurance that does not guarantee access and a lack of communication and coordination between the parents, the educational system and the health care system. Both the parents and the school personnel describe a system that is disjointed and disorganized. The lack of coordination of care and communication between school systems, parents, and the medical system is troubling and has significant consequences for accessing needed social and medical services. Moreover, the lack of access not only impacts the youth and their families, but it also negatively impacts the school staff, compromising their ability to provide the optimal educational experience for the youth.

Perhaps most critical for these individuals is the lack of preparation for the transition from a service system designed for children to an adult service system. Both parents and school personnel feel the system is disjointed and inadequate to meet the considerable needs of youth with disabilities as they transition from child health and mental health systems to adult health and mental health systems. Many are on a waitlist to receive Medicaid supportive services that could assure appropriate living conditions; however, it can take as long as seven years to receive such services. In addition, many parents report having only limited or no discussions with health providers about how the health needs of their youth will change as they get older, how to access the adult health care system and how to ensure adequate insurance that will provide access to health care services.

There are a number of important limitations to the study. The response rate to the survey was low and varied across sites. Discussions with school staff that the parents who did not respond to the survey had a lower literacy level and lower income than those who did respond. This was also confirmed by the difference between the overall demographics of the student/parent population compared to the survey respondents. For example, 45.8% of parents who responded to the survey had above a high school education whereas among the general parent population only 32% had above a high school education. Despite the low response rate, the survey validated and quantified many of the issues that were brought to light through the qualitative interviews. Another concern is that the study was conducted in one district in one state. Thus, there is limited generalizability of the findings to other settings. Nevertheless, this case study provides insights for the design of studies in other cities. Moreover, it provides ideas for future programmatic interventions that could end the cycle of disadvantage among youths with developmental disabilities.

In recent years there has been significant discussion at the national level around ensuring that individuals, especially those with chronic diseases have access to a medical home. Bodenheimer et al (23,24) for example discussed the need for realignment in the delivery system such that primary care physician's offices coordinate care with community resources, facilitate self-management support, and focus on providing care for both acute and chronic conditions. Indeed, the coordination of services across schools, medical providers and mental health providers and the development of integrated systems of supports were key recommendations that emerged from discussion with stakeholders. Medical homes can facilitate this process. Policy and programmatic efforts should focus on providing incentives for primary care physicians to serve as medical homes. One strategy, for example could be to increase Medicaid reimbursement to providers who serve as medical homes to individuals with severe and persistent disabilities.

A second idea that came out of our study was the need for school based health and mental health services specifically targeting children and youth with cognitive impairment and complex physical health issues. Indeed, Duval County does have a program entitled the Full Service School Program, that provides health and mental health services at schools (25). However, this program is focused on primary care and basic mental health services for the general school population. It excludes individuals with cognitive impairments and complex physical health programs. There is a large literature showing the benefit of school based health services including improving access to primary care, reducing health related school absences, improving nutrition and reducing obesity (26). However, there is a very limited experience in providing health and mental health services to behaviorally or medically complex children in the school setting (27). Based on our findings, significant needs could be addressed through a program that provides primary care, access to specialty medical and psychiatric consultation. This type of program could be provided through a combination of on-site care for basic primary care, care coordination, basic mental health, and dental. Telehealth could be utilized for consultations with mental and physical health specialists.

Lastly, we found that both parents and school staff were very anxious to have access to more complete and up-to-date information on the availability of services and programs for youth and young adults. The school personnel and parents said they had almost a complete lack of understanding of health and mental health services in the adult system of care for complex systems of care. As has been recommended in published guidelines,(28) a transition guide that is focused on services needed for youth and young adults with complex medical and developmental health issues is needed.

In our study we found that youth and young adults with cognitive impairment, behavioral problems and complex medical problems lack an integrated and coordinated system of supports for their health and health care, exacerbating already existing challenges for their parents and educational professionals. The lack of services conspires with the other threats encountered in impoverished communities to undermine the chances for a successful transition to adult care, adult career choices and independence. The transition to adulthood period is a time of great opportunity; however it is not well supported by health and educational services, resulting in the failure of many young adults to achieve their optimal level of independence. More ystems and services need to be put into place to help youth with cognitive impairment and complex medical programs to successfully transition to adulthood.

Acknowledgements

We would like to acknowledge the following entities and individuals for their support of this project and in the preparation of this manuscript:

- School Board of Duval County
- School administrators (principals, teachers, and nurses from: Alden Road, Mount Herman and Palm Avenue)
- Students from the ESE schools
- Southeast PEHSU and the Break the Cycle project
- Craig Rees and Victoria Chau (University of Florida MPH students)

References

[1] McPherson M, Arango P, Fox H, Lauver C, McManus M, Newacheck PW, Perrin JM, Shonkoff JP, Strickland B. A new definition of children with special health care needs. Pediatrics 1998;102:137-9.

[2] US Department of Health and Human Services, Health Resources and Services Administration, Maternal and Child Health Bureau. The National Survey of Children with Special Health Care Needs Chartbook 2005–2006. Accessed 2013 May 20. URL: http://mchb.hrsa.gov/cshcn05/

[3] Porterfield SL, McBride TD. The effect of poverty and caregiver education on perceived need and access to health services among children with special health care needs. Am J Public Health 2007;97(2):323-9.

[4] Lotstein DS, McPherson M, Strickland B, Newacheck PW. Transition planning for youth with special health care needs: Results from the National Survey of Children with Special Health Care Needs. Pediatrics 2005;115:1562-8.

[5] Lykens KA, Fulda KG, Bae S, Singh KP. Differences in risk factors for children with special health care needs (CSHCN) receiving needed specialty care by Socioeconomic Status. BMC Pediatr 2009;9:48.

[6] Arnett J. Emerging adulthood: The winding road from late teens through the twenties. Oxford: Oxford Univ Press, 2004.

[7] White P, Gallay L. Youth with special health care needs and disabilities in transition to adulthood. In: Osgood DW, Foster EM, Flanagan C, Ruth GR, eds. On your own

without a safety net: the transition to adulthood for vulnerable populations. Chicago, IL: Univ Chicago Press, 2005.

[8] Rosen DS, Blum RW, Britto M, Sawyer SM, Siegel DM. Transition to adult health care for adolescents and young adults with chronic conditions: Position paper of the society for adolescent medicine. J Adolesc Health 2003;33(4):309-11.

[9] McManus M, Fox H, O'Connor K, Chapman T, MacKinnon J. Pediatric perspectives on transitioning adolescents with special health needs to adult health care. Accessed 2013 Jan 10. URL: http://www.medicalhomeinfo.org/publications/transition.html #fact

[10] Peter NG, Forke CM, Ginsburg KR, Schwarz DF. Transition from pediatric to adult care: Internists' perspectives. Pediatrics 2009;123(2):417-23.

[11] Callahan ST, Cooper WO. Continuity of health insurance coverage among young adults with disabilities. Pediatrics 2007;119(6):1175-80.

[12] Fujiura GT, Yamaki K, Czechowicz S. Disability among ethnic and racial minorities in the United States. J Disabil Policy Stud 1998;9(2):111-30.

[13] Parish SL. Juggling and struggling: A preliminary work-life study of mothers with adolescents who have developmental disabilities. Ment Retard 2006;44(6):393-404.

[14] Coucouvanis K, Lakin CK, Prouty R, Webster A. Reduction continue in average daily populations of large state facilities; Nearly 70% decrease between 1980 and 2005. Ment Retard 2006;44(3)235-8.

[15] Kitchener M, Willmott M, Wong A, Harrington C. Home and community-based services: Medicaid research and demonstration waivers. San Francisco, CA: PAS Center Pers Assist Serv, 2006. Accessed 2013 May 24. URL:: http://www. pascenter.org/demo_waivers/

[16] The Kaiser Family Foundation: StateHealthFacts.org: Florida Home and Community Based Services. Accessed 2013 Jan 27. URL: http://www.statehealthfacts.org/ profileind.jsp?cat=4&sub=62&rgn=11

[17] US Census Bureau. Duval County, FL: U.S. Census Bureau, American fact finder. Accessed 2013 Jan 13. URL: http://factfinder.census.gov/home/saff/main. html?_lang=en

[18] Duval County Health Department. Center for Health Statistics Study, Health: Place Matters. Florida: DCHD Health, Policy and Evaluation Research Institute, 2008.

[19] Duval County Public Schools: Instructional research and accountability. Accessed 2013 May 24. URL: http://www.duvalschools.org/.

[20] Christ TW. A recursive approach to mixed methods research in longitudinal study of postsecondary education disability support services. J Mixed Methods Res 2007;1;226.

[21] Newacheck PW, Houtrow AJ, Romm DL, Kuhlthau KA, Bloom SR, Van Cleave JM, et al. The future of health insurance for children with special health care needs. Pediatrics 2009;123(5):940-7.

[22] Kenney MK, Kogan MD, Crall JJ. Parental perceptions of dental/oral health among children with special health care needs. Ambulat Pediatr 2008;8(5):312-20.

[23] Bodenheimer T, Lorig K, Holman H, Grumbach K. Patient self-management of chronic disease in primary care. JAMA 2002;288(19):2469-75.

[24] Bodenheimer T, Wagner EH, Grumbach K. Improving primary care for patients with chronic illness: The chronic care model, Part 2. JAMA 2002;288(15):1909-14.
[25] Duval County Public Schools: Instructional research and accountability. Accessed 2013 June 24. URL: http://www.duvalschools.org/static/students/services/full_service_schools.asp
[26] Lear JG. Health at school: A hidden health care system emerges from the shadows. Health Affairs 2007;26(2):409-19.
[27] Borenstein PE, Harvilchuck JD, Rosenthal BH, et al. Patterns of ICD-9 diagnosis using school-based clinics: Diagnostic categories by school level and gender. J Adolesc Health 1996;18(3):203-10.
[28] Stewart D. Transition to adult services for young people with disabilities: current evidence to guide future research. Dev Med Child Neurol 2009;51(Suppl 4):169-73.

In: Environmental Health
Editors: IL Rubin and J Merrick

ISBN: 978-1-62948-155-5

Chapter IX

Physical environment and student outcomes in green schools

Selen Okcu, PhD, Erica Ryherd, PhD* and Charlene Bayer, PhD
College of Architecture, Georgia Institute of Technology Atlanta,
George W Woodruff School of Mechanical Engineering,
Georgia Institute of Technology Atlanta
and Georgia Tech Research Institute, Atlanta, Georgia, US

The role of physical school environment on student health and education is becoming better understood. There is a growing body of literature that indicates improved physical environments in schools (e.g. indoor air quality, lighting and acoustic conditions) can enhance student outcomes. In parallel, the green building movement is centered around designing buildings, including schools, that are more sustainable in order to decrease energy consumption, minimize environmental impact and create healthier spaces for occupants. This paper synthesizes findings from both green design studies and school outcomes studies to provide a systematic evaluation of the potential impacts of green school design features on student outcomes. Three inter-related topics are covered in detail: (I)

* Correspondence: Erica Ryherd, PhD, Woodruff School of Mechanical Engineering, Georgia Institute of Technology Atlanta, Georgia 30332, United States. E-mail: erica.ryherd@me.gatech.edu.

overview of the "green" concept, including existing guidelines for "greening" schools, attitudes towards green schools and condition of the physical environments in non-green schools; (II) potential effects of the physical environment on school children, including documentation of national statistics and summary of findings from school research studies; (III) synthesis of findings, including a discussion of the knowledge gaps in the field of green school research and conclusions.

Introduction

Green schools aim to support wellbeing and performance outcomes of occupants by providing physical environments that are healthier, safer, more comfortable and productive (1,2). Some potential design features of green schools include: building envelope solutions effective in addressing mold and moisture problems, design and operation of ventilation systems effective in providing thermal control and avoiding pollutants and humidity, control of air pollutants; lighting solutions such as day lighting through windows, adequate lighting levels, good color rendering and use of blinds or external shading to avoid excessive direct and glazing sunlight, and acoustic solutions such as control of unwanted sources or use of absorptive treatments to reduce sound build-up. Massachusetts Technology Collaborative (MASSTECH) guidelines identified the three major missions of green schools as: cost less to operate, enhance learning and conserve resources (energy and water) (1). MASSTECH is a public development agency for renewable energy and the innovation economy that aims to maximize the benefits of clean energy. According to the Collaborative High Performance School Manual, green schools are expected to provide user and nature friendly environments that are healthy, safe, comfortable, secure, integrated with nature, efficient in material usage, energy and water consumption, maintenance and operation and adaptable to changing needs (3).

History of concepts

Valued building features for the occupants in schools have changed over time. In the United Kingdom, in 1931, the Hadow Report was released to address the needs of the school children between seven and eleven including physical and mental development, curriculum, school resources, staffing, internal organization of the school and the quality of physical environment in schools. The report highlighted the importance of sunlight and fresh air for children to promote good health, development and achievement outcomes and the necessity for school site selection in maximizing the sunlight and minimizing noise and dust from roads. The report also recommended cross ventilation in the school buildings (4). It was followed by another report by the Building Research Station (BRS) and the Medical Research Council (MRC) (5). This report highlighted the importance of adequate lighting levels, in particular maximized day lighting in the classrooms, for healthy eyesight, concentration levels and learning. As a result, during the 1950s, lighting was the driver for classroom design relative to ventilation, heating and acoustics. In the 1970s and 1980s, the green school concept started as a response to the energy conservation concerns (6). Later, in the 2000s, there was a trend to support school research to understand how the physical environment in schools impacts student and teacher outcomes. In 2002, President Bush's "No Child Left Behind Act of 2001" was passed (2). "Healthy, High-Performance Schools" subpart of the Act that was introduced by Sen. Hillary Clinton calls for national studies to assess the characteristics of unhealthy school conditions, the impact of unhealthy school conditions on student outcomes and provide recommendations and cost estimates necessary to improve environmentally unhealthy school buildings (7). In addition to national surveys, various research studies have been conducted to explore the association between school conditions and teacher and student outcomes. In the following sections, the findings of these studies are discussed.

School facilities in the United States

Several professional organizations and government agencies have started documenting the environmental conditions in US schools (e.g. GAO, NCES). Many US schools are old, lack proper maintenance and good design characteristics (that could potentially enhance health and performance outcomes). In 1999, the National Center for Educational Statistics (NCES) conducted the Fast Response Survey System (FRSS) and collected information on the physical condition of public schools based on nine different building features ranging from the condition of roofs to ventilation and air conditioning (8). According to the findings of this study, 50% of schools indicated that at least one of the nine building features at their school was not in adequate condition. The same study also found that 29% of the nation's schools had problems with heating, ventilation and air conditioning and 25% had problems with exterior walls, finishes, windows and doors. Another report by NCES showed that 28% of the public schools were built before 1950 and 45% between 1950 and 1969 (9).

These findings are particularly troubling, considering that school occupants spend a significant amount of time in these environments. According to the US Environmental Protection Agency estimates, a typical K-12 student spends 1,300 hours each year in the school building (10). Moreover, there is an increase in the number of school occupants. School enrollments increased 12% between the years 1990 and 1997 (11). According to US Census Bureau, in 2000, about 54 million K-12 students enrolled in schools were served by 5.9 million staff in those schools (10). To address these challenges, $28.6 billion was spent on school construction in 2001 (12), with approximately 58% going toward additions and modernizations. According to GAO estimates, approximately $112 billion is needed to bring schools up to good overall condition (13).

The profile of students in US public schools

The number of low-income students studying in US public schools has been constantly increasing since the 1990`s (14). In the South, low-income children have constituted a majority of the public school students for more than 40 years. The number of students of color (a.k.a. non-white students) in the public schools has also been increasing. According to national statistics, about 45% of the 49 million students studying in US public schools are low-income and/or students of color (15). According to 2010 Southern Education Foundation (SEF) research report, students of color now constitute a majority of the public schools in 11 states including Alabama, Arkansas, Florida, Georgia, Louisiana, Maryland, Mississippi, North Carolina, South Carolina, Tennessee and Texas (16). For example in Texas about two-thirds of all public school children are students of color. Overall, a majority of the students in US Public Schools, particularly in the South, is low-income and/ or students of color. Nationwide, low-income students and students of color perform, on average, below their peers. According to a 2007 National Assessment of Educational Progress (NAEP) test scores, low income students remain 20 to 30 points behind students not eligible for free and reduced lunch (14). For simple comparison, it is known that a ten point achievement gap difference in student performance scores equates to roughly a year`s worth learning for students. Some of the known factors leading to achievement gaps are lack of middle class "cultural capital", limited parental involvement and assistance with homework, lack of motivation, lack of resources, financial limitations and poor educational strategies adopted by schools (17-20). To close achievement gaps, that is the disparities in performance between all students and students from a background of social and economic disadvantage (students of color and low-income students), different school systems have been established. For example, students in the "Knowledge is Power Program" (KIPP) schools attend 9 hours of class a day on average with additional Saturday school and summer school (21). KIPP schools predominantly serve disadvantaged children. According to a 2008 KIPP report results, 81% of the KIPP students are low-income and more than 95% are African American or Latino.

Various resources indicate that these children are likely to study in schools with poor physical environments. The education of children from backgrounds of social and economic disadvantage costs more compared to other students (22). Therefore, schools by necessity spend a greater portion of their already limited funds on instruction and less on repairing buildings or buying or repairing equipment (23). For example, one study indicated that in an urban school district only about 3.5% of the budget is spent on facility maintenance and about 85% of this amount was for emergency repairs and the small amount of the remaining was spent on preventive maintenance (13). Unfortunately limited funding leads to deferred maintenance and escalated costs. Similarly, according to the results of a national study conducted by the National Center for Education Statistics (NCES) in 1999, schools with the highest concentrations of students already at a disadvantage are more likely to have less than adequate building features and unsatisfactory environmental conditions compared to schools with the lowest concentrations of disadvantaged students (24).

To summarize, the number of children from backgrounds of social and economic disadvantage in US public schools is increasing. Related with financial limitations, the schools with high percentages of disadvantaged children are likely to have very limited budgets for building maintenance and repair. Lack of maintenance is one of the major causes of poor conditions in US schools. In particular, children with no health coverage studying in these environments are high risk groups to be severely affected as the impact of poor physical conditions is highly unlikely to be detected or addressed.

What do authorities say about green schools?

Various research studies and articles support the concept of positive impacts of green schools on occupant outcomes. For example, research findings provide evidence showing that improved physical environmental qualities in school facilities, such as lighting, acoustics and air quality can enhance educational outcomes as discussed in the following sections (25-31). Additionally, the beliefs and experiences of school authorities generally support these research findings (32, 33). Turner Construction Company conducted a project called the "Green Building Market Barometer" (32). To obtain views on green schools, a survey was administered to 665 senior executives including building owners, developers, architects, engineers, corporate owner-occupants, consultants, and educational institutions. There was agreement among the executives about the positive impacts of green K-12 facilities. According to the findings of the project, more than 70% of the executives believe that green schools can enhance school districts ability to attract and retain teachers, help to improve student performance and reduce student absenteeism. A majority of the school authorities also believe that green schools can reduce the operational costs in the long run.

Heery International conducted research using focus groups and telephone interviews with staff such as teachers and principals from nine different areas in Oregon and Massachusetts (2, 33). The study analyzed the educators' opinions on the relationship between qualities of physical environments and educational outcomes. Almost all respondents believed there is a significant impact of physical conditions in schools on student learning and academic achievement. Eighty-seven percent also believed there is an impact on teacher retention. The educators rated lighting and climate control as the two most important design elements. Sufficient technology and flexibility of the space and windows were some of the other design elements rated as important. Most of the teachers desired to have the ability of controlling climate and decoration in the classrooms.

Cost effectiveness

Overall, green educational facilities are less expensive to maintain and lower in total costs due to reduced energy costs and efficient operation even though their initial costs are higher compared to traditional school buildings. According to the Turner Construction Company survey, school executives believe initial costs of green schools are higher, but more than 70% of the K-12 executives indicated their costs would be lower in the following 20 years (32). The same study also found that school districts and higher educational facilities mainly focus on initial construction costs and give less weight to total long term costs. One study conducted a cost analysis by considering 30 green schools (mostly LEED accredited) from 10 different states built between 2001 and 2006 (34). Four of those green schools in Georgia, Massachusetts and Oregon did not cost more than conventional design and the "green premium" (a.k.a. initial extra cost to build a green building compared to a conventional building) of other 24 schools ranged between 0.4% and 6.3%. Overall, green schools analyzed in the study mostly cost 1% to 2% more than schools with conventional design. This study also found that green schools with more efficient lighting, heating, cooling systems, maximized day lighting, sensor applications and effective insulation applications on the wall and roof systems consume about 33% less energy compared to schools with conventional design. Reduced energy use in green schools leads to annual emission reduction such as 1,200 pounds of nitrogen oxides, 1,300 pounds of sulfur oxide and 585,000 pounds of carbon dioxide. Furthermore, in these 30 green schools, an average of 32% reduction in water use was achieved. The Sustainable Building Industry Council (SBIC) estimated utilization of green school design features in new schools can save up to 30-40% and up to 20-30% building renovation projects (35). Also according to the SBIC estimates, an average middle school can save $500,000 over a 10-year period if day lighting strategies are utilized (36).

Guidelines for "better" schools

A variety of guides has been offered by different organizations in order to steer the efforts of school systems aiming to improve the performance and cost effectiveness of their facilities. These documents are intended to provide school districts a framework for design and construction of new buildings and for the operation and maintenance of existing buildings and protocols for the periodic inspection of health and safety conditions (37). Some of those guidelines were intended for the use of local school districts:

- Sustainable Design Guidelines by the Poudre School District (PSD), Colorado
- Best Practices Manual Volumes I, II, III by the Collaborative for High Performance Schools (CHPS), California
- Recommended Best Practices Manual, by Partnership for Resource-Efficient Schools, City of Seattle and Seattle Public Schools
- High Performance Guidelines: Triangle Region Public Facilities by Triangle J. Council of Governments (TJCoG), New York City
- Guidelines for Creating High Performance Green Buildings by Commonwealth of Pennsylvania, Pennsylvania
- High Performance Building Guidelines by City of New York, Department of Design and Construction, New York City
- Minnesota Sustainable Design Guide, by University of Minnesota (UM), Minnesota
- Health and Safety Guide for K-12 Schools, Washington
- OEHS Safe School Inspection Program, by Los Angeles Unified School District Office of Environmental Health and Safety, Los Angeles.

There are also guidelines intended for more global use such as:

- Indoor Air Quality (IAQ) Tools for Schools, by the US Environmental Protection Agency (EPA)
- Leadership in Energy and Environmental Design (LEED), by US Green Building Council
- Sustainable Schools, Innovative Design (ID)
- How to Build a Green School, by the Center for Health, Environment and Justice
- Advanced Energy Design Guide for K12 School Buildings, by the American Society of Heating Refrigerating and Air-Conditioning (ASHRAE)

Some of the guidelines developed by LEED, CHPS, University of Minnesota and TJCoG include a rating system. The US Green Building Council's LEED rating system is the most widely used sustainable design program throughout the United States (38). It was created in the late 1990s. Major LEED categories include sustainable sites, water, energy and atmosphere, materials and resources, indoor environmental quality and innovation and design. The guideline emphasizes energy and resource efficiency, low or no toxicity products and materials, site disturbance, reducing atmospheric degradation and enhancing indoor environmental quality. LEED's rating system offers four levels of certification. They are: certified, silver, gold and platinum.

School children and growing health concerns

There is a growing concern about the school children's health. The origin of these concerns is poor physical environments such as acoustic conditions, air quality and lighting levels. The US Government Accountability Office (GAO) estimated that there are more than 14 million children studying in physical school environments that may be a threat to their health (13).

For example, many students experience hearing problems. A national study surveyed the hearing health of 6,166 children aged between 6-19 years by conducting hearing tests at 0.5 to 8kHz (39). Among those children approximately 15% had low-frequency or high frequency hearing loss at least 16dB hearing level in one or both ears, about 7% had low frequency loss of at least 16dB hearing level, and approximately 13% had high frequency loss at least 16dB hearing level. A GAO study estimated that approximately 22,000 U.S. schools attended by about 11 million students had unsatisfactory acoustic applications for noise control (13). In the same study, 28% of the schools reported "unsatisfactory" or "very unsatisfactory" sound environments. The study findings indicated that young children with normal hearing require significantly higher signal-to noise ratios, compared to adults, to achieve similar speech acquisition levels. Therefore, they are the most severely affected population group from high noise levels and poor acoustic conditions such as excessive reverberation (40). Children's ability to understand in noise improves over time and reaches adult performance levels during the teen years (41). A detailed study found the performance of nine year olds in recognition of sentences in noise was poorer than 11, 15 and 17 year olds (42).

In addition to hearing loss, a significant number of the school children are believed to have vision problems. One study obtained health information from 5,141 children via the 1988 Medical Expenditure Panel Survey and provided national estimates. It is believed that nationwide more than 25% of the 52.6 million children between 6-18 years old had corrective lenses (43). However, these data may underestimate the need for vision correction. Because of socio economic factors, a significant student population who need a vision correction might not have been evaluated or diagnosed with a vision problem (1). Furthermore, findings indicate that seasonal reductions in the amount of daylight can induce depression and result in seasonal affective disorder (SAD) (44). Even though it has not been studied widely among school children, it is believed that less learning can be expected among children experiencing SAD (1). For the treatment of SAD, bright light from electric lighting systems is preferred (45).

Another concern is that many children experience asthma related absences. According to study estimates in 2003, among four million children who experienced asthma, students missed 12.8 million school days because of their asthma (46). Another study in 2007 found that 5.6 million school children aged between 5-17 years old suffered from asthma (47).

According to the GAO, more than 8.3 million students attend 15,000 schools with unsatisfactory indoor air quality (13). The same study also found that in 19% of the schools participating in the study, indoor air quality and heating conditions are "unsatisfactory" or "very unsatisfactory" and same ratio was 27% for ventilation.

Physical school environments and school children outcomes

There is a growing body of literature that compares the outcomes of school children studying in schools with poor and improved physical conditions (e.g. newer, recently renovated). One study examined the impact of physical conditions of school buildings on student achievement and student behavior in 47 rural high schools in Virginia (48). Physical conditions of the school buildings were assessed based on the Commonwealth Assessment of the Physical Environment. The Test of Academic Proficiency for 11th graders was used to assess student performance. Student behavior was determined by comparing of the number of expulsions, suspensions and incidences of violence/ substance abuse to the total number of students. Socio-economic status of students (the ratio of the number of students not on free and reduced lunch to total number of students) was used as a covariate to control the effects of socio-economic status on student achievement and behavior. According to the findings of the study, student achievement was higher in schools with better physical conditions. The study also found that science achievement was higher in schools with better lab facilities.

Another study analyzed the impact of the physical environment on 4th and 6th grade student achievement, health, attendance and behavior. The outcomes of 280 students studying in the oldest and the newest school facilities in a rural Tennessee county school district were compared (25). The older school building from the late 1930`s contained fluorescent lighting, a coal fired furnace and several air conditioning units. The new/modern school contained fluorescent lighting, electric heating and air conditioning. Unlike the old school building, in the new school additional efforts were made to control acoustic quality, coordinate color schemes and use up-to-date furniture. After controlling for effects of socio-economic status of the students, the study found that students attending the newer school had significantly higher achievement in math, reading, listening and language than those in the older school facility. Additionally, attendance was higher and reduced major health problems and disciplinary actions were reported in the newer school.

Another study indicated that students had more positive attitudes about their school in a newer school compared to students in an older school (49). Cheng also explored the impact of perceived physical environment on student affective performance (50). The students rated the physical conditions in the school based on eleven items. For the assessment of student affective performance, measures considered were self-concept, attitudes towards peers, attitudes toward teachers, self-efficacy of learning, feeling of homework overload, and intention to drop out of school. More than 21,600 students from 190 schools participated in the study. The study showed that higher quality physical environments were positively correlated with all measures of student affective performance except self-concept.

Maxwell studied the impact of renovations as a measure of building condition on student performance in 21 elementary schools in New York City School District (31). Schools that underwent a major renovation within the past ten years were considered as "recently renovated". Pupil Evaluation Program (PEP) test results of 3rd and 6th graders from the 1982- 1983 through the 1996-1997 school years were analyzed. It was found that math scores were significantly and positively correlated to the percentage of students attending recently renovated schools. Similarly, there was an increase in the percentage of the students who have limited proficiency in English. The study also underlined the importance of the timing of the renovations as exposure to noise and experiencing the chaotic environment in a building undergoing major renovation may temporarily result in decreased achievement outcomes.

Chan compared the impact of the presence and absence of air conditioning, carpet, fluorescent lighting and interior pastel coloring on student achievement outcomes measured by scores on the Iowa Test of Basic Skills (51). The study reviewed the test scores of 8th grade student from 191 public schools in Georgia. The percentage of student participation in paid lunch program was used as a covariate to control the variance in achievement outcomes due to the effect of socio-economic status. The study also minimized the variance in achievement outcomes due to teacher qualification, teacher-student ratio and per student expenditure by including only standard public schools in the state. The findings of the study indicated student's vocabulary scores were higher in air-conditioned schools compared to non-air conditioned schools. The impacts of other physical school features were not significant.

Lighting

Evidence from school lighting research indicates that improved school lighting can enhance both visual (e.g. healthy vision) and non-visual (e.g. achievement) student outcomes. Most of the previous findings indicate that lighting conditions in the classrooms have important non-visual effects on students. However, the amount of knowledge about the visual effect of lighting on school children remains limited. Heschong Mahone Group studied the effects of day lighting from windows or skylights on student performance (52). Math and reading test scores of 21,000 students from three school districts (California, Washington, and Colorado) were compared. Window tint and size, presence and type of skylights and amount of expected day light were the measures used to define the lighting conditions in the classrooms. Students with more day lighting processed 20% faster on math tests and 26% on reading tests in one year. Those with greater window areas progressed 15% faster in math and 23% faster in reading. In the classrooms with a well designed skylight, diffusing the daylight throughout the room, student achievement scores improved 19-20% faster than those without a skylight. Additionally, students in classrooms with operable windows progressed about 7-8% faster than those with non-operable windows.

Grandgaard compared the impact of white walls and cool-white fluorescent lighting (standard lighting) with light blue walls and full spectrum lighting (28). Off-task behavior and blood pressure were measured for 11 six year old children. A decrease of 22% in off-task behaviors and 9% in student blood pressure was observed in the conditions with light blue walls and full spectrum lighting.

Ott studied the impact of lighting on hyperactivity and learning disabilities (53). The behavior of first grade students in two windowless classrooms with standard cool-white fluorescent lighting were compared with the behavior of the students in the two windowless classrooms with full-spectrum fluorescent lighting that is very similar to natural daylight. In the former condition, fixtures with solid plastic diffusers were used. In the latter lead foil shields were wrapped around each end of the tubes to reduce ultraviolet radiation exposure caused by the fluorescent lamps. Study findings indicated a dramatic improvement in the hyperactive behaviors of children in a couple of months after the lighting interventions were applied. However the study did not mention any information about the diagnosis of hyperactive disorder among those children. According to observers, the students were less nervous, paid more attention to the teacher and overall performance was better.

Another study looked at the relationship between different lighting conditions in schools and dental health, growth, development, attendance and academic achievement outcomes (54). Outcomes of students exposed to four different types of artificial lighting were compared. Those were full-spectrum fluorescent lamps, full-spectrum fluorescent lamps with ultraviolet light supplements, cool-white fluorescent lamps, and high-pressure sodium vapor lamps. The study was conducted in two phases. A total of 327 and 233 fourth grade students were surveyed from five schools in 1987 and in 1989 respectively. No significant differences were found in the sex, age, or daily nutrition among those students. The students under full-spectrum fluorescent lamps with ultraviolet supplements developed fewer dental cavities and had better attendance, achievement and growth and development than did students under other lighting conditions. Students under the high-pressure sodium vapor lamps had the slowest growth and development rates as well as the poorest levels of attendance and achievement. This is mainly because full-spectrum lighting with ultraviolet supplements is closest to natural day light. Natural day light provides the spectral energy distribution for different biological functions including skin reddening, vitamin D synthesis, pigmentation of the skin, bilirubin degradation as well as reduction of dental cavities (55).

Noise

Exposure to high noise levels and poor classroom acoustic conditions can negatively affect school children's educational (e.g. reading, language acquisition and cognitive abilities) and health outcomes. A literature review by Schneider found that there is good evidence showing the link between acoustic conditions in schools and student outcomes such as spelling and reading abilities, behavior, attention, concentration, blood pressure, feelings of helplessness and a lack of persistence on tasks (56). Another literature review by McGuffey indicated that unwanted noise adversely affects student performance and desirable music can help to improve student test scores in reading comprehension (57). Earthman and Lemasters found evidence from the reviewed literature showing that excessive noise levels are associated with student stress, dissatisfaction with the classroom, and lower achievement (58).

The findings of empirical studies show that the impact of noise on student outcomes can vary depending on the type of noise, task being performed, etc. Several studies have looked at the impact of anthropogenic environmental noise such as aircraft and traffic noise. For example, Green et al. studied the relationship between the percent of students reading below grade level and noise exposure level in all elementary schools in Brooklyn and Queens in New York (59). Noise Exposure Forecast contours for New York City airports were used to assign noise exposure scores to schools. This study indicated that the percent of students reading below grade level is negatively correlated with noise levels. Another study compared the cognitive performance and health of 451 elementary school children attending 10 schools in high aircraft noise areas (Leq>63dBA) with the children attending 10 schools in lower aircraft noise areas (Leq< 57dBA) (60). This study, after controlling for school and individual level confounding factors, found a significant association between exposure to aircraft noise and impaired reading on difficult items and raised annoyance. Stansfeld et al conducted a cross-national study and explored the association between aircraft and traffic noise and student cognitive abilities and health (61). About 2800 elementary school children attending schools around major airports in Spain, UK and Netherlands participated in the study. Noise exposure levels were predicted from noise contour maps, modeling and on-site measurements. Standardized tests and questionnaires were administered in the classrooms to measure cognitive and health outcomes. Exposure to chronic aircraft noise was significantly associated with impairment of reading comprehension, recognition memory and annoyance after controlling for school and individual level confounding factors. Exposure to traffic noise was associated with episodic memory (conceptual recall and information recall) and annoyance. None of the two noise types affected sustained attention, self-reported health and overall mental health.

Evans and Maxwell studied the impact of chronic exposure to noise on language acquisition and reading deficits of first and second grade students with normal hearing (27). It was found that exposure to chronic aircraft noise rather than interference during the reading was related to reading deficits and impaired speech perception after controlling for mother`s education. This study also highlighted the mediator effect of language acquisition on the relationship between noise exposure and reading deficits.

Christie and Glickman explored the impact of noise levels on students` intellectual performance (26). One hundred and fifty-six 1st, 3rd and 5th grade school children from a public school in Ohio were asked to perform a type of intelligence test known as Standard Progressive Matrices. Students were exposed to tape-recorded classroom noise during the experiment. The study findings indicated that noise affected children differently. While female students performed better in quiet (40dBA) rather than noisy (70dBA) environment, male students were able to solve more complex matrix problems in noisy environments. Children`s performance on the intellectual task increased with age.

Indoor air quality and thermal comfort

Schools with improved indoor air quality and increased ventilation rates can improve student performance and health outcomes. Wargocki et al. conducted an experimental study and compared the outcomes of the students exposed to different air quality conditions in two classrooms (62). In one of those classrooms, average air temperatures were reduced from 23.6°C to 20°C and outdoor air supply rates were increased from 5.2 to 9.6 liters per second (l/s) per person. Each condition was tested for a one-week period. During lessons, children were asked to perform eight different tasks developed to represent their schoolwork (e.g. reading, mathematics). They were also asked to complete a survey each week and indicate the intensity of the building-related symptoms. Findings showed that increased ventilation rate significantly increased student performance in doing addition, multiplication and number checking. Reduced temperature significantly increased student performance in subtraction and reading. Reduced temperature at increased ventilation rate significantly increased student performance in a test of logical thinking.

Overall, the study suggested increasing ventilation rates from 5.4 to 9.6 l/s per person and decreasing temperatures from 23.6°C to 20°C would potentially improve the performance of school children.

Smedje et al (63) investigated the impact of improving the ventilation systems in schools on children's health outcomes. They administered questionnaires to 1,476 primary and secondary school children in randomly selected 39 schools during 1993-1995. Ventilation rates were examined in 100 classrooms. With the installation of new ventilation systems in 12% of the classrooms, air-exchange rates were increased and relative humidity and concentration of several airborne pollutants were reduced. The incidence of asthmatic symptoms was less common among 143 children who attended classrooms with the new ventilation systems in 1995 compared to 1993.

Shendell et al (64) explored the association between student absences and a new air quality measure which is the difference between indoor minus outdoor carbon dioxide levels (d CO_2). Five minute CO_2 samples from 409 traditional and 25 portable classrooms from 22 schools were collected in six school districts in Washington and Idaho. Student attendance was retrieved from school records. When school and individual level confounding factors were controlled, it was found that a 1,000 parts-per-million increase in d CO_2 was associated with 10 to 20 % increase in student absences that is suggested to be an indirect measure of student performance. Furthermore, 45% of the classrooms had short-term indoor CO_2 concentrations above 1,000 ppm. Note that ASHRAE recommends that indoor CO_2 levels should not exceed the outdoor concentration by more than 600ppm (65). Since outdoor CO2 levels generally range between 300-400ppm, indoor CO_2 levels exceeding 1,000ppm are considered to indicate inadequate ventilation.

Another study investigated the levels of cat (Fel d I) and dog allergens (Can f I) by collecting dust samples from tables, chairs and floors in 29 classrooms in four Swedish schools during 6 week period (66). Higher Can f I and Fel d I levels were found on chairs than on tables and floors that was potentially brought to school by students and teachers, on their clothes. Furthermore the allergen levels were high enough to sensitize children and induce asthma in most children allergic to cats and dogs.

Bates and Mahaffy measured the levels of airborne and surface fungi, bacteria and dust mite allergen as well as indoor and outdoor temperature and relative humidity in 13 carpeted classrooms in six Florida schools (67). Students and teachers were asked to report their health complaints or allergy symptoms. In each classroom, air samplers were set up 24 inches above the carpet and 8 hours of data were collected as well as additional air samples for mold spores, carpet dust samples for mold and mite allergen and information about visible mold and odor. After eight hour data collection, carpets in the classrooms were cleaned by a cleaning firm. Higher levels of mite allergens were found in the classrooms with higher relative humidity. Health complaints, musty odors and visible mold growth were associated with high indoor relative humidity over 69%. No health complaints occurred in schools with newer chillers and extended air conditioning times. Carpet cleaning significantly reduced the levels of dust mite allergen and mold spores in the carpet dust.

Koskinen et al compared respiratory symptoms among 229 children who are exposed to visible mold growth on interior walls of daycare centers with those who were not exposed to mold growth (68). Parents were asked to record health symptoms of their children (3-7 years old). Students exposed to visible mold experienced high risks of sore throat, purulent and non-purulent nasal discharge, nasal congestion, hoarseness and common cold compared to non-exposed children.

Parallel to these research findings, the Institute of Medicine (IOM) issued two reports on the association between excessive moisture/dampness, mold growth and the development of asthma and exacerbation of asthma symptoms (e.g. nasal congesting, sneezing, runny or itchy noise) among children and adults. In 2000, IOM`s released the first report on "Cleaning the air: asthma and indoor air exposure". This report investigated the level of scientific evidence linking the indoor air components with the causes of asthma and suggesting the strategies for exposure mitigation and prevention (69). According to the IOM study findings, even though it has been known that changing the environment of an asthmatic individual can reduce symptoms; little data are available to conclude which specific strategies can be considered as public health measures, because of the many confounding factors that can also impact asthma incidence and severity, including genetic factors and other socioeconomic differences. Therefore, until the development of a more fundamental understanding is achieved, the study suggested the use of multifaceted strategies such as removal of particulate allergens both from home and schools (e.g. reduction of cat dander, use of HEPA filters, high ventilation rates and air cleaning systems).

In 2004, IOM released the second report on "Damp indoor space and health" and reported the level of scientific evidence based on four categories: sufficient evidence of a causal relationship, sufficient evidence of an association, limited or suggestive evidence of an association and inadequate/insufficient evidence to determine whether an association exists

(70). The committee found sufficient evidence that indicate exposure to damp indoor environments adversely impacts several measures of respiratory health (such as upper respiratory tract symptoms; asthma symptoms in sensitized asthmatic persons; and cough, wheeze and hypersensitivity pneumonitis in susceptible persons).

Improving student outcomes through green school applications

Although there is a good deal of evidence to suggest that the physical environment impacts student outcomes, there are not many rigorous studies that have specifically examined the impact of green schools on student outcomes. However, as described below, interviews with architects and reports of school authorities indicate that people perceive that green schools can enhance student outcomes. Different from the "better" or "newer" school concept for specific systems (e.g. lighting, acoustics, air quality) considered in the reviewed studies above, green schools pursue a holistic approach in enhancing the overall physical environment in schools which includes multiple systems.

One study analyzed student academic achievement outcomes in three green schools (34). In Ash Creek Intermediate School in Oregon, a LEED certified school, green school design applications lead to 15% reduction in student absenteeism. Similarly, in Clearview Elementary School in Pennsylvania, a LEED Gold school, there was 19% increase in average Student Oral Reading Fluency Scores (DIBELS). Moreover, in Third Creek Elementary School in North Carolina, a LEED Gold school, green design applications helped to improve student learning and test scores and there was 12% decrease in missed teacher workdays.

Miners reported the results of a post occupancy evaluation (POE) study conducted by the officials from the Columbia-Brazoria Independent School District to assess the impacts of green design features applied in West Brazos Junior High School in Texas (71). The high school is a LEED certified public school and some of the green design applications in the school include special glazing on windows to improve thermal efficiency, shading devices that help to distribute light further into learning spaces, the use of green cleaning products and the applications of reflective roofing and paving system to reduce accumulation of heat at certain zones (a.k.a. heat island effect). Student performance scores based on the Texas Assessment of Knowledge and Skills (TAKS) in the new high school were compared with the TAKS in the old building. The results indicated 5 point improvement in reading scores, 4 point in math and 7 point improvement in social studies. Additionally, in the new high school, referrals also decreased significantly.

Edwards compared the achievement outcomes of the students attending green and non-green schools in UK with similar social, geographic and size characteristics (6). In total, 54 green schools built between 1975 and 1995 were considered. Some of the green design features applied in these schools include solar design, maximized natural light and ventilation by adopting shallow floor depths, thermal flywheel technology, maximized use of local building materials to reduce embodied energy and maximized exposure to nature. Several of the identified problems in these green schools include high windows that are difficult to open, inadequate solar shading and excessive reflection on computer screens and air temperatures that were too high and too low during winter and summer times due to interventions to maximize day light. Student achievement outcomes were assessed by SAT scores, absenteeism rates, teacher turnover rates, teaching days lost due to illness and evaluations provided by Office for Standards in Education (Ofsted) inspectors. The study findings indicated that green schools that gave priority to daylight and natural ventilation generally outperformed other schools in the same school district. The student outcomes in older passive solar schools were not as high as the student outcomes in recently constructed green schools due to conflict between ventilation and energy efficiency. The study suggested the consideration of both the needs of the learning and energy conservation when employing green strategies to maximize day light levels and solar heating.

Outcome measures and methods used to assess student outcomes

In the school research studies described above, occupant health and performance outcomes are quantified with the use of different measures. The impact of the physical environment on student health has been investigated by relevant outcome measures such as the respiratory system (e.g. asthma, cold/flu, difficulty in breathing), cardiovascular system (e.g. heart rate and blood pressure), dermatologic (e.g. itching, dry-irritated skin), hearing, vision, dental and behavioral (e.g. concentration/attention) health outcomes, headaches, fatigue, eye-nose-throat irritations, stress, annoyance and growth-development outcomes. Different methods were used to obtain health data such as health symptom questionnaires or, diary checklists completed by parents.

Student performance was quantified with the use of various outcome measures such as math, language and reading test scores, recognition and episodic memory outcomes and absenteeism. Different methods were used to obtain student performance data. Observational methods were used to measure the level of student attention to teacher, student-teacher interaction and student concentration. Student absenteeism was obtained from school attendance records. Very few studies developed their own student performance tests that were parallel to school content. Another study used comprehensive lists of student self-report items while assessing student affective performance. Other studies used different educational inspection tools including Screening Instrument for Targeting Education Risk (SIFTER) completed by the teachers and the reports provided by Office for Standards in Education inspectors. However, mainly standardized reading and achievement tests were used. These tests are conducted on a yearly basis throughout the states which mostly include reading, language and math. The examples of those standardized tests include SAT Reasoning Test, Texas Assessment of Knowledge and Skills (TAKS), California Achievement Tests (CAT), Comprehensive test of basic skills (CTBS), Iowa Test of Basic Skills (ITBS), Tests of Achievement and Proficiency (TAP), Wisconsin Student Assessment System (WSAS) and Wyoming Comprehensive Assessment Systems (WyCAS).

Control measures

Various confounding factors were considered in the reviewed studies exploring the link between physical school environments and children outcomes. Some of those commonly considered control measures include socioeconomic status, age differences among students, primary language spoken at home, mother's education level, attendance rate, percentage of gifted enrollment, percentage of mobility rate, percentage of the students receiving free lunch, student background data, demographics such as race, gender and health problems such as hearing loss. Particularly, age was considered to be critical for some studies as compared to teenagers young children have higher relative volume of breathing per weight, actively growing organs and tissues (1) and developing language and cognitive skills (42).

Condition of school buildings and some design features were also considered as confounding factors. Some of those control variables include age, size (number of enrolled students) and type of the school, geographic location (e.g. rural, urban) and financial abilities of the school. These variables can be critical in making direct comparisons between schools as some of the conditions in schools are not standardized.

Conclusion

A significant number of US schools are old, and lack proper maintenance and good design characteristics that potentially enhance health and performance outcomes of children. In US public schools, the number of children who come from backgrounds of social and economic disadvantage is growing. The cost of education for this group of children is higher. Therefore, typically the schools with high percentages of children at a disadvantage because of social and economic background have limited budgets for building maintenance and repair. Lack of building maintenance is one of the well known causes of poor physical conditions in US schools. Empirical appreciation as well as various studies do indicate that children at social and economic disadvantage are more likely to study in schools with poor/unsatisfactory physical environments.

The number of studies exploring the impact of the physical environment in schools on student health and education outcomes is increasing. The researchers` evaluation of the existing literature is as follows:

1. Scientific research examining the impact of desirable physical conditions in new or recently renovated schools on student outcomes by comparing the student outcomes from traditional or older schools remains limited. In general, the results indicate that student outcomes improved with enhanced physical conditions in schools.
2. There are few studies that looked at the combined effects of multiple building systems on occupant outcomes. The majority of the studies looked at the overall impact of design (e.g. newer versus older schools) and association between student outcomes and particular qualities of physical environments such as air quality, thermal, lighting and acoustic conditions.
3. There appears to be an association between student learning and health outcomes, and indoor air quality and thermal comfort in the classrooms. These studies generally showed that increased ventilation

rates, control of excessive moisture, visible molds and air temperature can protect the health of occupants and improve comfort and student learning in schools.

4. Improved lighting conditions can improve student academic performance, behavior, health, growth and development outcomes. A growing body of evidence from non-school research also suggests the significant impact of seasonal reduction in the amount of day light on health outcomes such as depression. There are few studies on the visual effects of light on students. It is believed that impact of poor lighting conditions can be more dramatic for the students who need eye correction as they can experience high visual discomfort even during the normal lighting conditions provided in the classrooms.
5. Finally, there is a significant impact of reduced noise levels in schools on student achievement outcomes. Young children are more severely affected by the high noise levels related with their developing speech discrimination skills. However, there are not many studies on the relationship between reverberation time or other physical acoustic properties on student performance.

Despite research exploring the impact of poor and improved school conditions on school occupants, we still do not know much about the specific impacts of green schools on occupant outcomes. There are only a few evidence-based design (EBD) studies about the overall effects of green schools on occupant outcomes. At this point, it can be concluded that there is a need for well-designed EBD studies to highlight the benefits of green schools for its occupants.

Acknowledgements

This work has been partially supported by Southeast Pediatric Environmental Health Specialty Unit (SE PEHSU) Student Award. We would like to acknowledge collaborations with Break the Cycle Environmental Health Disparities project leaders Dr. Leslie Rubin, Janice Nodvin and others who helped facilitate this work. This chapter is a revised and adapted version of an earlier publication: Okcu S, Ryherd E, Bayer C. The role of physical environment on student health and education in green schools. Rev Environ Health. 2011;26(3):169-79.

References

[1] Committee to Review and Assess the Health and Productivity Benefits of Green Schools. Green schools: Attributes for health and learning. Washington, DC: National Research Council, 2007.

[2] Bosh S. Identifying relevant variables for understanding how school facilities affect educational outcomes. Atlanta, GA: Georgia Institute Technology, 2003.

[3] Collaborative for High Performance Schools (CHPS). Best Practices Manual. 2004. Accessed 2013 May 25. URL: www.chps.net

[4] The Hadow Report: The primary school. London: HMSO,1931.

[5] Building Research Station (BRS), Medical Research Council (MRC). Research Report. RIBAJ 1947;54:1-7.

[6] Edwards BW. Environmental design and educational performance, with particular reference to green schools in Hampshire and Essex. Res Educ 2006;76:14-32.

[7] No Child Left Behind Act of 2001. Healthy and high performance schools. 2002. Accessed 2013 May 25. URL: http://www.healthyschools.org/HHPS_Act_2002.pdf.

[8] National Center for Education Statistics (NCES). Condition of America's public school facilities: 1999. Washington, DC: US Department Education, 2000.

[9] National Center for Education Statistics (NCES). How old are America's public schools?. 1999. Accessed 2013 May 25. URL: http://www.ed.gov/NCES/pubs.

[10] SPM. Green building survey shows increased interest in K-12 realm. Sch Plann Manage 2006;45(4):G3-G8.

[11] U.S. General Accounting Office (GAO). School facilities: Construction expenditures have grown significantly in recent years. Washington, DC: GAO, 2000.

[12] Agron J. Bucking the trend: The 28th annual official education construction report. 2002. Accessed 2013 May 25. URL: http://asumag.com/mag /university_bucking_trend_th/index.html

[13] US General Accounting Office (GAO). Conditions of America`s schools: America`s schools report differing conditions. Washington, DC: GAO, 1995.

[14] Southern Education Foundation (SEF). A new majority: Low income students in the South`s public schools. Research report. Atlanta, GA: Southern Education Foundation, 2007. Accessed 2013 May 25. URL: www.southerneducation.org.

[15] Healthy Schools Network. Sick Schools 2009: America`s continuing environmental health crisis for children. Research Report. Albany, NY: Healthy Schools Network, 2009.

[16] Southern Education Foundation (SEF). A new diverse majority: Students of color in the South`s public schools. Research report. Atlanta, GA: Southern Education Foundation. Accessed 2013 May 25. URL: www.southerneducation.org.

[17] Annette Lareau. Social class differences in family-school relationships: The impact of cultural capital. Sociol Educ 1987;60(2):73-85.

[18] Jencks C, Phillips M. America's next achievement test: Closing the black-white test score gap. Am Prospect 1998;40:44-53.

[19] Hart B, Risley T. Meaningful differences in the everyday experiences of young children. Baltimore, MD: Paul H Brookes, 1995.

[20] Suarez-Orozco C, Suarez-Orozco M. Transformations: immigration, family life, and achievement motivation among Latino adolescents. Palo Alto, CA: Stanford University Press, 1995.
[21] 2008 KIPP Report Card. Accessed 2013 May 25. URL: www.kipp.org/reportcard/2008.
[22] US General Accounting Offi ce (GAO). School facilities: America ' s schools report differing conditions. Report to Congressional Requester. Washington, DC: US General Accounting Office, 1996.
[23] National Center for Education Statistics (NCES). Disparities in public school district spending 1989 – 90. Washington, DC: US Department Education, 1995.
[24] National Center for Education Statistics (NCES). Condition of America's public schools facilities: 1999. Statistical analysis report. Washington, DC: US Department Education, 2000.
[25] Bowers JH, Burkett CW, editors. Relationship of student achievement and characteristics in two selected school facility environmental settings. The 64th Council of Educational Facility Planners, International Conference, Alberta, Canada, 1987.
[26] Christie DJ, Glickman CD. The effects of classroom noise on children: evidence for sex differences. Psychol Schools 1980;17:405-8.
[27] Evans GW, Maxwell L. Chronic noise exposure and reading deficits – the mediating effects of language acquisition. Environ Behav 1997;29:638-56.
[28] Grangaard EM. Color and light effects on learning. Association for Childhood Education International Study Conference and Exhibition, Washington, DC, 1995.
[29] Green GH. The effect of indoor relative humidity on absenteeism and colds in schools. ASHRAE Trans 1974;80:131-41.
[30] Heschong L, Wright R, Okura S. Day lighting impacts on human performance in school. J Illum Eng Soc 2002;31:101-14.
[31] Maxwell LE. School building renovation and student performance: one district ' s experience. Scottsdale, AZ: Council of Educational Facility Planers International, 1999.
[32] Executives See Educational Value of Green Building. IEQ Strategies 2006;19:6-8.
[33] Heery International. Millennium school 2000. Atlanta, GA: Heery International, 2000.
[34] Kats G. Greening America ' s schools: cost and benefi ts. A Capital E Report, October 2006. Accessed 2013 May 25. URL: www.cap-e.com.
[35] Sustainable Building Industry Council (SBIC). High performance school buildings, 2001. Accessed 2013 May 25. URL: www.sbicouncil.org.
[36] Reicher D. Nature ' s design rules. Learn Design 2000;9:16-8.
[37] Bosh S. Characterization of guidance documents for creating high performance schools. Atlanta, GA: Georgia Institute Technology, 2002.
[38] Kellert SR. Designing healthy schools. Independent Sch 2005;65:58-61.
[39] Niskar AS, Kieszak SM, Holmes A, Esteban E, Rubin C, et al. The prevalence of noise-induced hearing loss among children 6-19 years of age: the 3rd National Health and Nutrition Examination Survey, 1988 – 94 – United States. Am J Epidemiol 1998;147:296.

[40] Nabelek AK, Robinson PK. Monaural and binaural speech-perception in reverberation for listeners of various ages. J Acoust Soc Am 1982;71:1242-8.
[41] Nilsson M, Soli SD, Sullivan JA. Development of the hearing in noise test for the measurement of speech reception thresholds in quiet and in noise. J Acoust Soc Am 1994;95:1085-99.
[42] Elliott LL. Performance of children aged 9 to 17 years on a test of speech-intelligibility in noise using sentence material with controlled word predictability. J Acoust Soc Am 1979;66:651-3.
[43] Kemper AR, Bruckman D, Freed GL. Prevalence and distribution of corrective lenses among school-age children. Optometry Vision Sci 2004;81:7-10.
[44] Rosenthal NE. Winter blues, seasonal affective disorder. What is it and how to overcome it ? New York: Guilford, 1998.
[45] Rosenthal N, Sack D, Parry B, Mendelsom W, Tamarkin L, et al. Seasonal affective disorder and phototherapy. Ann NY Acad Sci 1985;453:260.
[46] Akinbami LJ. The state of childhood asthma, United States, 1980 – 2005. Hyattsville, MD: National Center for Health Statistics, 2006.
[47] American Lung Association, Epidemiology and Statistics Unit, Research and Program Services. Trends in asthma morbidity and mortality. Washington, DC: American Lung Association, 2009.
[48] Cash CS. Building condition and student achievement and behavior. Blacksburg, Virginia: Virginia Polytechnic Institute and State University, 1993.
[49] Chan TC. A comparative study of pupil attitudes toward new and old school buildings. Greenville, SC: South Carolina School District of Greenville County, 1982.
[50] Cheng YC. Classroom environment and student affective performance: An effective profi le. J Exp Educ 1994;62:221-39.
[51] Chan TC. Physical environment and middle grade achievement. Greenville, SC: South Carolina School District of Greenville County, 1980.
[52] Heshong Mahone Group. Day lighting in schools: an investigation into the relationship between day lighting and human performance. Fair Oaks, CA: Heschong Mahone Group, 1999.
[53] Otto JN. Infl uence of fl uorescent lights on hyperactivity and learning disabilities. J Learn Disabil 1976;9:22-7.
[54] Hathaway WE. Effects of school lighting on physical development and school performance. J Educ Res 1995;88:228-42.
[55] Hathaway WE, Hargreaves JA, Thompson GW, Novitsky D. A study into the effects of light on children of elementary school age, February 1992. Accessed 2013 May 25. URL: http://www.naturallighting.com/articles_effects_of_lighting_on_school_children.php.
[56] Schneider M. Do school facilities affect academic outcomes? Washington, DC: National Clearinghouse Educational Facilities, 2002.
[57] McGuffey CW. Facilities. In: Walberg HJ, ed. Improving educational standards and productivity. Berkeley, CA: McCutchan, 1982:237-81.

[58] Earthman GI, Lemasters L, editors. Where children learn: a discussion of how a facility affects learning. The Virginia Educational Facility Planners Annual Meeting, Blacksburg, VA, 1998.
[59] Green KB, Pasternack BS, Shore RE. Effects of aircraft noise on hearing ability of school-age children. Arch Environ Health 1982;37:284-9.
[60] Haines MM, Stanfeld SA, Brentnall S. The West London school study: the effects of chronic aircraft noise exposure on child health. Psychol Med 2001;3:1385-96.
[61] Stansfeld SA, Berglund B, Clark C, Lopez-Barrio I, Fischer P, et al.; RANCH Study Team. Aircraft and road traffi c noise and children ' s cognition and health: a cross-national study. Lancet 2005;365:1908-9.
[62] Wargocki P, Wyon DP, Matysiak B, Irgens S. The effects of classroom air temperature and outdoor air supply rate on the performance of school work by children. Proceedings of Indoor Air 2005 (10th International Conference on Indoor Air Quality and Climate), Beijing, China, 2005:368-72.
[63] Smedje G, Norback D. New ventilation systems at selected schools in Sweden – effects on asthma and exposure. Arch Environ Health 2000;55:18-25.
[64] Shendell DG, Prill P, Fisk WJ, Apte MG, Blake O, et al. Associations between classrooms CO 2 concentrations and student attendance in Washington and Idaho. Indoor Air 2004;14:333-41.
[65] American Society of Heating Refrigerating and Air-Conditioning Engineers (ASHRAE). Ventilation for acceptable indoor air quality. ANSI/ASHRAE standard 62, 1989.
[66] Munir AK, Einarsson R, Schou C, Drebork SK. Allergens in school dust. J Allergy Clin Immunol 1993;91:1067-74.
[67] Bates JM, Mahaffy DJ. Relationship of reported allergy symptoms, relative humidity and airborne biologicals in thirteen Florida classrooms. Proceedings of Indoor Air ' 96 (7th International Conference on Indoor Air Quality and Climate), Nagoya, Japan, 1996:551-6.
[68] Koskinen O, Husman T, Hyvarinen A, Reponen T, Nevalainen A. Respiratory symptoms and infections among children in a daycare center with mold problems. Indoor Air 1995;5:3-9.
[69] Institute of Medicine (IOM). Clearing the air: asthma and indoor air exposures. Washington, DC: National Academies Press (NAP), 2000.
[70] Institute of Medicine (IOM). Damp indoor air spaces and health. Washington, DC: National Academies Press (NAP), 2004.
[71] Miners Z. From smart construction to smart learning. (Facilities and Construction) District Administration, 2008;April:16 – 7. Accessed 2013 May 25. URL: http://www.districtadministration.com/viewarticle.aspx ? articleid = 1539.

In: Environmental Health
Editors: IL Rubin and J Merrick

ISBN: 978-1-62948-155-5

Chapter X

Guidelines for better school siting

Robert J Geller, MD*[*], *Rebecca Watts Hull, MS, Denise R Grabowski, AICP, LEED AP and Katherine Moore, AICP

Emory University School of Medicine, Atlanta;
Mothers & Others for Clean Air, American Lung Association in Georgia, Smyrna, Georgia;
Georgia Conservancy, Atlanta, Georgia; Symbioscity, Savannah, Georgia, US

The benefits of screening potential school sites for environmental hazards and the special vulnerability of children to environmental pollutants and toxics are well understood. Less apparent, and less widely discussed until recently, are other health concerns associated with school location, such as ready access to healthy food options and opportunities for safe walking and bicycling. In recent years, a variety of stakeholders interested in children's wellness in school settings have expressed concerns about the trend to place larger schools further away from the communities they serve. In some cases, this trend can worsen environmental health risks, such as air pollution, and reduce opportunities for physical activity, with disproportionate impacts on communities already burdened with multiple

[*] Correspondence: Robert J Geller, MD, Emory University Department of Pediatrics, 50 Hurt Plaza SE Suite 600, Atlanta GA 30303, United States. E-mail: rgeller@emory.edu.

stressors. This paper describes a project in Georgia that helps school system decision-makers more fully integrate children's environmental health and wellness concerns into all aspects of school siting planning, through a training curriculum based on the U.S. Environmental Protection Agency's 2011 School Siting Guidelines. Two training modules have been developed. Preliminary feedback indicates that the process itself—convening a diverse set of stakeholders with different expertise related to children's environmental health, community planning, and school system planning—results in new partnerships and the consideration of other perspectives. Better integration of municipal and school planning and improved understanding of environmental health considerations can improve public health outcomes for the entire community.

Introduction

In recent decades, in communities throughout the United States, school systems have made decisions resulting in larger and larger school facilities, often located on the outskirts of the communities they serve. In 1953, 3,113 schools served 863,761 students in Georgia (1). By 2011, the population of school-age students had grown to more than 1.6 million, but the number of schools serving them actually decreased, to 2,246 (2). At the same time, in response to financial pressures, changing demographics, and other factors, many older, community-centric school buildings have been closed. In 2002, this "mega-school sprawl" trend began to receive greater attention, as reports raised questions both about the loss of historic, small schools and the implications of larger and larger facilities for student learning and health (3).

The trend toward larger schools located further from residential community centers has had many unintended consequences, including worsened traffic congestion and "sprawl" development patterns (4,5). In some cases, this trend also has resulted in a loss of connectedness between schools and communities, due to the increased separation between housing and school facilities (6,7). In addition to the community-scale impacts of recent trends in school siting, the public health community has become increasingly concerned about the consequences of these trends on the health and wellness of children and youth. Potential negative health impacts include reduced physical activity as a result of limited opportunities to walk or bike to school, and increased exposure to vehicular air pollution as a result of increased car and diesel bus use and greater proximity to multi-lane, busy roadways.

In 2011, the US Environmental Protection Agency (EPA) published the first set of comprehensive, voluntary guidelines for school siting (8). The guidelines were developed as the result of a 2007 directive from the U.S. Congress that specifically requires model guidelines for the siting of school facilities (9). Congress directed EPA to consider the following in its development of the guidelines: the special vulnerability of children to environmental hazards; ways in which school siting decisions influence student and staff travel options; energy efficiency of school buildings; and opportunities for using school buildings as emergency shelters.

The first directive—to consider "the special vulnerability of children"—speaks to the importance of school siting as one strategy for breaking the cycle of environmental health disparities. Children are more vulnerable than adults to many common environmental hazards (10), and they may spend as much as 33 hours per week in school, including travel (11). As a result, addressing a school's relationship to the larger community it serves and the built environment of which the school campus is a part is a critical component of breaking the cycle.

This paper describes a project in Georgia that is using EPA's new voluntary school siting guidelines to help decision-makers improve health through smarter planning for school facilities, with the goal of addressing one aspect of the environmental health disparities cycle.

Training workshop

The most expertly crafted guidelines do not meet their goals if they fail to reach the intended target audience. School siting is a very complex process. As described above, the consequences of school siting decisions have significant impact on the health and wellness of children. As a result, three Georgia-based nonprofit organizations—Georgia Conservancy, Inc., Mothers & Others for Clean Air, and US Green Building Council-Georgia Chapter, with financial support from U.S. EPA and the Arthur M. Blank Family Foundation, developed a training workshop for educational leaders and other school siting decision-makers based on EPA's guidelines.

The project team, comprised of staff from the organizations listed above and EPA Region IV, assembled an advisory committee of experts representing the fields of pediatrics, environmental health, community planning, green building, parent involvement, state education policy, and school district planning. Members of the advisory committee represent nationally respected and highly knowledgeable organizations including the Centers for Disease Control & Prevention (CDC), Southeast Pediatric Environmental Health Specialty Unit at Emory University (PEHSU), Georgia Institute of Technology, EPA's Office of Children's Health, Emory University's Rollins School of Public Health, Georgia Chapter of the American Academy of Pediatrics, National Trust for Historic Preservation, and Georgia Parent-Teacher Association (PTA).

An experienced training curriculum writer with expertise in the subject matter was hired to plan and facilitate the workshops and to craft two versions of the school siting training curriculum (three-hour and one-hour lengths). The training modules were developed utilizing EPA's School Siting Guidelines, the National Trust for Historic Preservation's Helping Johnny Walk to School, and other related materials.The advisory committee provided input during the development the training modules through two interactive workshops, followed by a "beta pilot" of the draft three-hour training module.

During the first workshop, held in November 2011, the advisory committee was introduced to the key elements in the aforementioned EPA guidelines, the principal policy recommendations of Helping Johnny Walk to School, and current policies and funding formulas in Georgia that influence school size and siting decision-making. Committee members worked in small, interdisciplinary groups to suggest priority considerations for decision-makers regarding the extensive list of issues related to school siting.

At a second workshop in January 2012, a first draft of the three-hour module was presented to the advisory team for its consideration and feedback was gathered. Following additional revisions, the second draft was presented to the advisory team as well as several representatives from the target audiences as a "beta pilot" in April 2012. Additional feedback was incorporated into the draft and the final versions were completed in June 2012.

Our findings

The process described above resulted in the development of two school siting training modules: a three-hour, interactive workshop suitable for pre-conference events, and a one-hour overview of school siting issues better suited for presentation-style opportunities.

In addition to introducing participants to key environmental and public health concerns in layperson's language, the longer training module includes problem-solving exercises. These hands-on exercises allow participants to share their own knowledge and experiences with the school siting processes, and to discuss and reflect on ways in which their perspectives were broadened as a result of the workshop. An exercise showing a theoretical community faced with a school siting decision is shared early in the workshop. Towards the end, participants are given the opportunity to wrestle with tough trade-offs and apply new information and perspectives to the exercise (see figure 1).

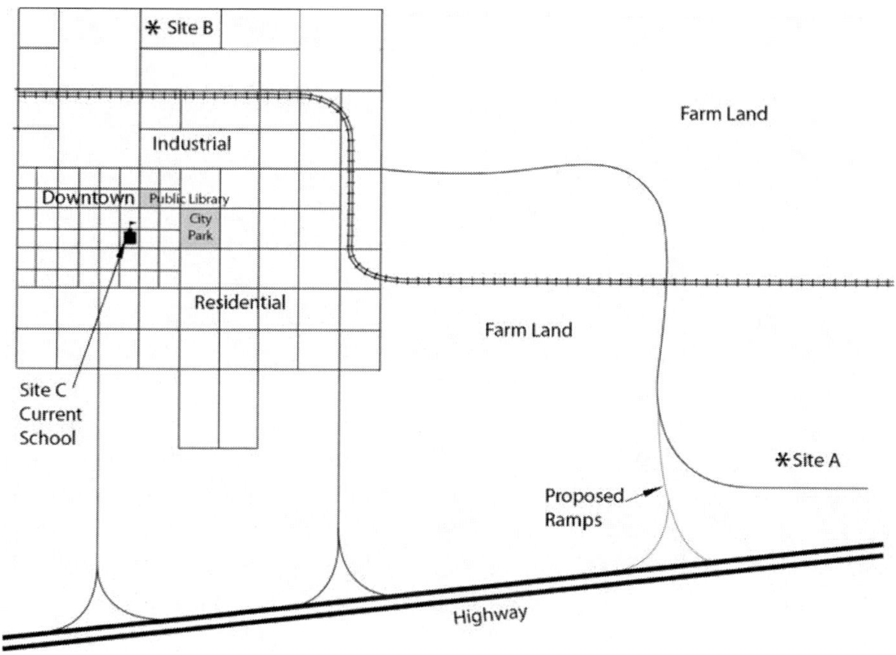

Figure 1. Credit: Anthony Viers.

While the training module contains references to Georgia-specific examples and an introduction to Georgia school siting policies, it was designed to be transferable to other states. The critical issues surrounding school siting are not unique to Georgia and have similar impacts on environment, public health and community characteristics across the nation. Several other states have expressed an interest in adapting the training module.

Portions of the three-hour training module are described here, with an emphasis on children's health considerations. Following an introduction to the origin and rationale for EPA's School Siting Guidelines and the process used to develop the training module, the workshop emphasizes the importance of meaningful public involvement, desirable school site attributes, and avoiding environmental hazards or risk factors.

A. School siting principles

The EPA guidelines are built upon four key principles:

- Safe and healthy school environments are integral components of the educational process.
- The environmental review process should be rigorous, thorough and well documented and include substantive and ongoing meaningful public involvement.
- The school siting process should consider the environmental health and safety of the entire community, including disadvantaged and underserved populations.
- Schools should be located in environments that contribute to the livability, sustainability and public health of neighborhoods and communities.

While these principles intuitively make sense, community environmental health and safety and its importance to student learning often have not been considered early in decisions about school location, outside of avoiding very clear hazards. EPA's guidelines encourage school systems to integrate health and wellness concerns into every phase of decision-making about school facilities.

B. Meaningful public involvement

While school districts usually have sole authority to make school siting decisions, those decisions have impacts far beyond the school campus boundaries. As a result, decision-making processes about school siting should involve a broad group of stakeholders. EPA's guidelines emphasize the importance of meaningful public involvement, and they expand the definition of "public," in a school context, beyond the group of stakeholders often consulted on school system decisions. The training curriculum introduces participants to the types of stakeholders to engage, discusses why it is critical to do so, and then provides an opportunity for participants to examine a school siting decision from many different points of view.

A common characteristic of building collaboration lies in the ability to see issues from multiple perspectives. Viewing issues from multiple perspectives almost always rules out the "simple solution" and reveals underlying values, motivation, root causes, common goals and unintended consequences (12). The curriculum invites participants to consider differences in perspective among four types of stakeholders that might be expected to have a stake in evaluating school site alternatives: school system representatives (administrators), school users (teachers, students, parents and families), local government (staff and elected officials), and community members.

After a brief discussion of likely differences and commonalities among the four stakeholder groups, participants break into small groups and explore how members in each of the four groups might respond to a theoretical school siting case involving a new elementary school. The exercise forces workshop participants to systematically view the complex decisions involved from the perspectives of others.

A number of participants engaging in this exercise during the "beta pilot" provided feedback that their understanding of the "best" choice for the theoretical school site and the inherent trade-offs changed quite a bit as a result of working through the perspective of each stakeholder group. In addition, participants also noted that their opinions were influenced as a result of the comments and opinions expressed by others in their work group.

C. Desirable school site attributes (And avoiding undesirable characteristics and hazards)

In addition to a participatory process and the need to consider children's health and wellness at every step, EPA's guidelines and the training module emphasize what kinds of characteristics are desirable in a school site, when feasible. Many, if not most, state departments of education require an environmental assessment of some sort before a school system can get approval for a new facility. These required assessments focus largely on avoiding existing hazards and, to some extent, potential hazards from nearby sources of pollutants (13). EPA's guidelines and this training module, however, emphasize that school siting decisions involve much more than the presence or absence of hazards. Planned well, they can contribute to school and community health and wellness; even community vitality. Some of the desirable attributes, as well as hazards to avoid, described in the curriculum are:

Health benefits of healthy, high performance schools: EPA's guidelines recommend that school systems consider the option of "green" building design. Green building approaches can result in improved indoor environments, including better air quality, improved acoustics, and access to natural light and views. Green or "high performance" school design has been demonstrated to result in improved health and comfort among staff and students as well as improvements in academic performance (14). Because cost is a highly significant consideration in school system decision-making, the training module discusses the relative increased cost of construction compared to the cost savings of green buildings (primarily due to reduced energy use) as well as health and learning benefits (15).

Community-centered schools with "safe routes": The alarming rise in childhood obesity rates in the United States has resulted in increased attention among schools, communities and policy-makers to opportunities for young people to engage in physical activity. Obesity among children and adolescents in the U.S. more than tripled between 1980 and 2008; in 2008, more than one-third of U.S. children aged 6-19 were overweight or obese (16,17). Public health officials recognize that "active transportation"—walking or biking to get where we need to go, including school—offers a good way to ensure regular physical activity. However, the number of children and youth walking or bicycling to school has dropped significantly, from 47.7% of students in 1969 usually walking or biking to school, compared with just 12.7% of K–8 students in 2009 (18).

The training curriculum reviews obesity trends in the U.S., along with the steep decline in the number of children walking or bicycling to schools, and uses aerial images of alternative schools and school/community configurations to illustrate how school siting and minimum acreage requirements affect children's opportunities to build physical activity into their day. Participants have another opportunity to work with the options and trade-offs with another theoretical school siting scenario.

Avoiding—and not generating-- air pollution: Based on input from the advisory team and the team's awareness that roadway air pollution is a significant health concern in Georgia's urban areas, the training curriculum includes more detail about air pollution concerns and their relation to school siting decisions than it does for some other environmental hazards. The EPA guidelines also give special attention to air pollution concerns, a highly prevalent concern in urban and suburban areas. Air pollution and its relation to school siting is important to consider not only because it has adverse health effects on children and youth, similar to other environmental hazards including groundwater contamination, but also because school siting decisions actually influence air pollution concentrations. As a result, as decision-makers consider a siting alternative along a busy roadway, in addition to measuring or modeling potential air pollution exposures along that road, they also should consider (and model, if possible) the likely impacts on air pollution concentrations (and related exposures).

The training module describes the interplay of children's health, physical activity and air pollution exposures resulting from a school siting decision's implications for transportation. Safe walking and bicycling options not only provide opportunities for physical activity, they also mean fewer cars and/or buses, and less exhaust (see figure 2).

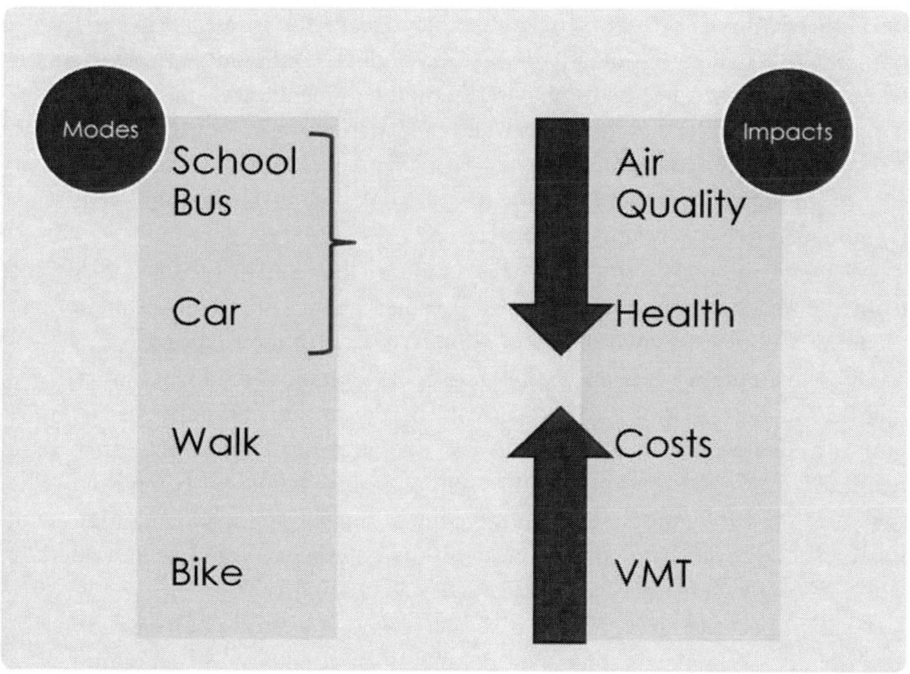

Figure 2. Credit: Symbioscity.

Discussion

Over the next several years, the project team anticipates delivering the completed training module to school board members, superintendents, school system planners, city and county planners, and nonprofit organizations throughout the State of Georgia. Interested groups and EPA staff in other states are planning to adapt the module for their use. However, some immediate changes in school siting practices and approaches have occurred as a result of the project itself, and its process.

Coincidentally, this project was underway at a time of significant change in the Atlanta Public School (APS) district. In late 2011, APS publicly released a proposal for redistricting schools and closing some facilities in order to accommodate changing demographics and to reduce costs. The original proposal had very significant consequences for the ability of children in some communities to safely walk or bicycle to a neighborhood school.

Learning about this project, a number of parent leaders concerned about the redistricting plans contacted the project team for information on EPA's guidelines and their rationale for recommending community-centric schools where feasible. In addition to resources and recommendations in the guidelines regarding walkable, neighborhood schools, parents also began talking with APS leadership about community input and involvement—a clear priority in EPA's guidelines, but often not within school system leadership—and were emboldened to expect a more inclusive and transparent process.

Second, the purposefully diverse nature of the advisory team led to some team members reaching out to staff in other agencies and departments and planning school siting in a more collaborative way for the first time.

For example, one team member, who leads construction planning for a middle Georgia school system known for its strong commitment to green building, decided to meet with the county planning and zoning office as a result of her participation on the school siting advisory committee. She reviewed the EPA guidelines with the county planner and made a commitment for the school system to collaborate with the county and other stakeholders in siting five new elementary schools planned in coming years.

The project also has sparked discussion among a number of Atlanta-based air quality scientists about a potential pilot study investigating the effectiveness of vegetative barriers to mitigate schools already sited on very busy roadways.

A more collaborative siting process is believed to be critical to achieving more desirable outcomes—improving student and community environmental health and potentially breaking the cycle of health disparities in some communities. As a result, the project team plans to adapt the training modules described in this paper specifically for parents and community members, modifying the curriculum slightly for a lay audience. These versions will then be made available broadly through local Parent Teacher Associations and other school community networks.

Acknowledgements

This project was funded by a grant from the U.S. Environmental Protection Agency, Source Reduction Assistance Grant Program (Multi-Regions Projects). The authors are grateful for the financial support that made the project possible.

We also would like to acknowledge and thank the members of Southeast PEHSU for including a presentation of the school siting training curriculum project at the 2012 Break the Cycle conference. This chapter is a revised and adapted version of an earlier publication: Geller R, Hull RW, Grabowski DR, Moore K. Breaking the cycle through better school siting: A collaborative project to facilitate the effective use of EPA's guidelines with Georgia's educational leaders. Int J Disabil Hum Dev 2013;12(4).

References

[1] Georgia Department of Education. Annual reports of the Department of Education to the General Assembly of the State of Georgia. Atlanta, GA: Georgia Department Education, 1954.

[2] Georgia Department of Education. 2010-2011 Report Card. Accessed 2013 Jun 01. URL: http://www.doe.k12.ga.us/Pages/By-School.aspx

[3] Beaumont CE, Pianca EG. Why Johnny can't walk to school. Washington, DC: National Trust Historic Preservation, 2002.

[4] US Environmental Protection Agency. Travel and environmental implications of school siting. Washington, DC: EPA 231-R-03-004, 2003.

[5] Michigan Land Use Institute. Hard lessons: Causes and consequences of Michigan's school construction boom. Beulah, MI: Michigan Land Use Institute, 2003.

[6] Salveson D. Good schools-good neighborhoods. Chapel Hill, NC: University North Carolina, Center Urban Regional Studies, 2003.

[7] Kuhlman, R. Helping Johnny walk to school: Policy recommendations for removing barriers to community-centric schools. Washington, DC: National Trust Historic Preservation, 2011.

[8] US Environmental Protection Agency. School siting guidelines. Washington, DC: USEPA, 2011.

[9] United States Congress. Energy Security and Independence Act, 2007.

[10] Frumkin H. Introduction. In: Frumkin H, Geller R, Rubin IL, Nodvin J. Safe and healthy school environments. New York: Oxford University Press, 2006:3-10.

[11] Hofferth SL, Sandberg JF. How American children spend their time. Ann Arbor, MI: Population Studies Center, University of Michigan, Report #00-458, 2000.

[12] Williams R, Cowie G, Olson S. Institute for Georgia Environmental Leadership, May 2009.

[13] Georgia Department of Education, Facilities Service Unit. Guidelines for risk hazard assessment of educational facility sites, 160-5-4.16(a)5.

[14] Eley C. High-performance school buildings. In: Frumkin H, Geller R, Rubin IL, Nodvin J. Safe and healthy school environments. New York: Oxford University Press, 2006:331-350.

[15] Stokes B. Bibb County Schools, personal communication, 2012.

[16] Hedley AA, Ogden CL, Johnson C, Caroll M, Curtin L, Flegal K. Prevalence of overweight and obesity among US children, adolescents, and adults, 1999-2002. JAMA 2004;291(23):2847–50.

[17] Ogden CL. Prevalence of high body mass index in US children and adolescents, 2007–2008. JAMA 2010;303(3):242–9.

[18] McDonald NC, Brown AL, Marchetti LM, Pedroso MS. US school travel, 2009: An assessment of trends. Am J Prev Med 2011;41(2):146-51.

Section 3: The Community

In: Environmental Health
Editors: IL Rubin and J Merrick
ISBN: 978-1-62948-155-5

Chapter XI

A media reflection on the cycle of disadvantage and disability

Rasaan Jones, MPH *
Morehouse School of Medicine, Atlanta, Georgia, US

This chapter describes a project using the "Media Madness/Getting the Rap on the Media" model and a conceptual framework developed by the Jane Fonda Center. The purpose of the project was to provide an educational program to reduce the influence of the media on the sexual attitudes and behaviors of youth. The project and intervention is an enhancement of a media curriculum, focused on the subject of teen dating violence, self-worth, and the Cycle of Disadvantage and Disability (CDD) model. This project has three components: 1) Research on teen dating violence, 2) Exploration of the relationship between the CDD, self-worth, teen dating violence, and the media and 3) Educating, engaging, and helping adolescents to develop healthy relationships. This paper documents the first phase of the Media Refection Project: risk assessment, research, methodology, and intervention development. The next phase is implementation and evaluation.

* Correspondence: Rasaan Jones, MPH, Centers for Disease Control and Prevention, 1600 Clifton Road NE, Atlanta, GA 303291193 Chris Lane SW, Mableton, GA 30126 United States. E-mail: imn1@cdc.gov.

Introduction

The concept of self-worth is important in adolescent development (1). The cycle of disadvantage and disability (CDD) model examines self-worth as a critical component. The Institute for the Study of Disadvantage and Disability in Atlanta, Georgia notes that children who are born into and who grow up in circumstances of social and economic disadvantage are more likely to have low self-worth, and this may lead to behaviors that perpetuate the cycle (see figure 1) (2). This project "A media reflection" focused on two important factors contributing to self-worth as it relates to adolescent health, teenage behaviors, and mass media influences: teen dating violence and risky unhealthy behaviors. A media reflection had two phases; phase one included the overall development of the project while phase two was the implementation and evaluation of the intervention. This paper uses the Jane Fonda Center's Media Madness Model to document the essential steps in raising awareness among adolescents pertaining to teen dating violence and negative media messages. The two main objectives of the project were to:

- explore the influence of media on relationship abuse among teens and the impact of teen dating violence on adolescents
- identify strategies that help teens learn to resist harmful media messages that may promote teen dating violence and unhealthy relationships.

A literature review was conducted on the topic of teen dating violence. We used these data to develop a conceptual framework and a logic model as a basis for an educational activity (intervention). The logic model describes an innovative intervention to help teens resist negative media messages and understand teen dating violence.

The paper's intended audience is program staff involved in planning interventions and programs aimed at improving reproductive health outcomes of teens while considering the impact of mass media. The critical premise of this project is: engaging, educating, and empowering youth to break the cycle of violent unhealthy relationships will assist in breaking the cycle of disadvantage and disability.

The media reflection project was completed at the Jane Fonda Center at Emory School of Medicine. It was an extension of the media madness curriculum, developed and currently being implemented by the Center. The curriculum is aimed at educating teens to learn to resist negative media messages about sex and sexual behaviors. The project's goal is to reach 25,000 middle school youth throughout Georgia by training health educators, community stakeholders, and youth leaders to implement the curriculum.

Teen dating violence is a relevant and pertinent issue that allows us to explore teen behaviors that may be associated and influenced by the mass media (3). From a public health perspective, teen dating violence presents a good subject for developing a curriculum enhancement that can be used within the media madness project format.

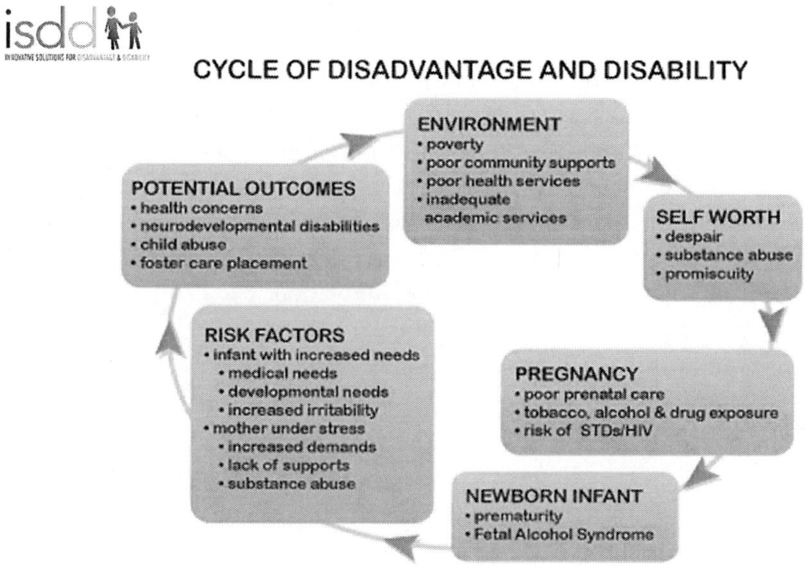

Figure 1. Cycle of disadvantage and disability.

Figure 1 depicts the cycle of disadvantage and disability (CDD). Within the CDD, the environment phase can pose many threats to the welfare and health of a community (2). The environment phase, at the beginning of the model, refers to issues of poverty, low socioeconomic status, poor health services and community supports. It is important to note that children, minorities, and disabled people are most at risk of being affected by negative aspects of their environment (4).

The negative influences and behaviors shaped by these environmental factors are sometimes reflected in an adolescent's own self-worth, the next phase within the CDD model. The self-worth phase is important and critical to child development, as low self-worth often elevates risk and exposure to substance abuse and can lead to pregnancy, which is the next phase in the cycle (2). Intimate partner violence among adolescents is also associated with increased risk of substance use, as well as unhealthy weight control, sexual risk behaviors, pregnancy, and suicide (5). The risk factors and potential outcomes for infants born under such circumstances perpetuate the CDD model.

Low self-worth, despair, substance abuse, and promiscuity are all unhealthy feelings and behaviors that lead to risky sexual behaviors and pregnancy. Within the CDD, low self-worth is a central risk factor for risky behaviors. However, one may ask the question: Is low self-worth a result of how people feel about themselves or a reflection of how they see themselves? Gerbner's cultivation theory argues that viewers who watch a great deal of television are more likely to imitate what they see and perceive as reality (6). This paper partly addresses this bridge between CDD, mass media and adolescent development using this theory of cultivation

Literature review

The Centers for Disease Control and Prevention (CDC) uses a four-step approach to address public health problems, including teen dating violence (7). Step one is to define the problem by gathering and studying relevant data. Step two includes identifying risks and protective factors. The final two steps involve developing and testing prevention strategies and adopting interventions on a wide scale. This framework was used in conducting this project. The first two steps were completed by conducting a literature review. We then developed a conceptual framework and an intervention but the intervention on a wide scale was beyond the scope of the project.

We conducted a literature review on abusive relationships and on media associations with teen dating violence. Additionally, information from dating violence resource centers and the center for Media Literacy was used to assemble facts pertaining to basic aspects of teen dating violence. This provided critical inputs for a definition of teen dating violence and the conceptual model used to develop the curriculum enhancement.

Research on media influences and adolescent behaviors was retrieved from journal articles and databases. This represented the second step of the CDC approach; identifying risks and protective factors. The Medline search used the keywords: teen dating violence, teen dating abuse, teen relationships and violence and media and teen dating violence. The literature search enabled us to identify risk factors associated with teen dating violence as well as risk factors for other behaviors associated with teenage relationship abuse.

Informational sources were arranged into a manual, separated by: 1) facts, 2) findings, 3) resources and 4) journal research articles. An annotated bibliography was also attached for a quick reference guide. Developing and testing prevention strategies (step three), followed by widespread adoption (step four), are the final steps in the CDC approach (7). Children in low income communities are more likely to be vulnerable to media influence (4). In many cases, the media can represent a type of "role model" for adolescents who may not have sufficient parental guidance or responsible adults around them, to counteract these influences. Therefore, the media reflection project focused on an intervention for adolescents in these communities.

Four major guidelines were used as criteria for developing the intervention. First, there had to be an interactive portion of the activity; second the intervention had to be fun and able to keep the youths' attention; third the activities had to be age-sensitive regarding the subject material; and fourth there had to be a section for youth to express their opinions. These standards must be tested for effectiveness among the target population in the second phase of the media reflection project, which is the implementation and evaluation phase.

The intervention was developed by using information from the literature review, the four guidelines mentioned above, and the conceptual framework or logic model (see figure 2). The first activity is a quick lecture introducing the topic of teen dating violence to the audience followed by a "myths and facts" worksheet. The inputs for this activity come directly from the findings of the literature review. The next activity includes an "agree or disagree" theme, which is an interactive activity using discussion topics about teen dating violence. This activity examines how youth need to think critically about common perceptions on abuse and violence within the adolescent community. Lastly is the creation of a teen dating bill of rights which provides an empowering aspect of the intervention.

The logic model and media madness formats guided the development of an innovative way to educate youth. This model also aided in developing the project outline, design, and variables required to successfully accomplish the specific short and long-term project goals.

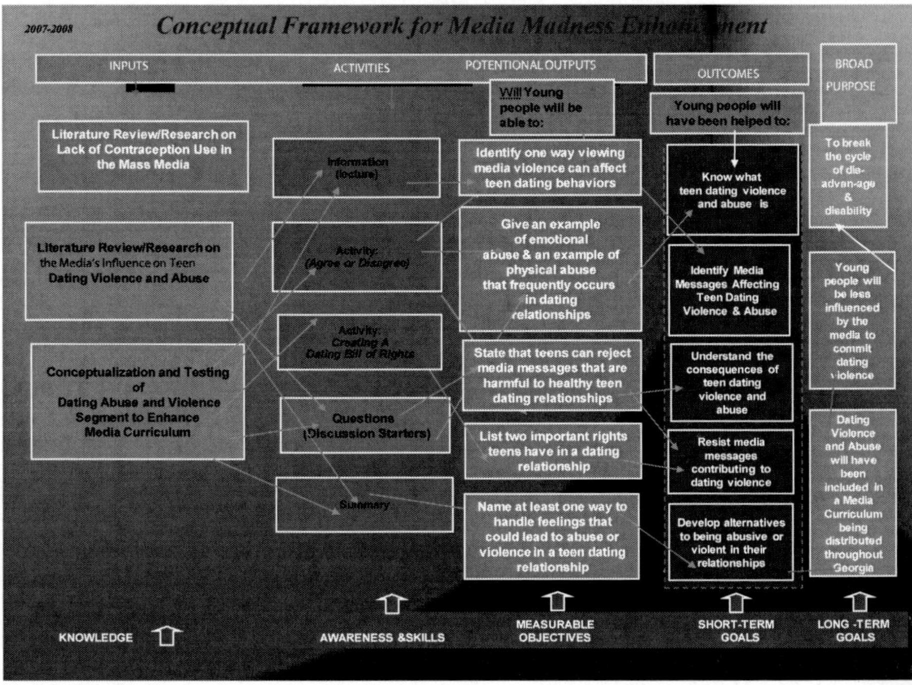

Figure 2. Conceptual framework.

Findings

Dating violence is controlling, abusive, and aggressive behavior in a romantic relationship. It can include verbal, emotional, physical, or sexual abuse, or a combination of these. The National Center for Victims of Crime explains that, "both boys and girls can be victims of dating violence, but boys and girls abuse their partners in different ways". They state, "boys physically injure girls more, thus are more likely to punch their partners, and boys also force their partners to participate in unwanted sexual activity." Meanwhile, "girls are more likely to yell, threaten to hurt themselves, pinch, slap, scratch, or kick their partners" (8).

Some teen victims experience violence occasionally, while others are abused more often, sometimes daily. However, "one in five teens in a serious relationship report having been hit, slapped, or pushed by a partner (8)." Using a national sample, the CDC reported that the average prevalence of dating violence for high school and college students is 22% and 32%, respectively (9). Many studies indicate that, as a dating relationship becomes more serious, the potential for and nature of violent behavior escalates (10). More importantly, studies have also concluded that abuse from dating partners can significantly elevate risk for a board range of serious public health concerns and burdens.

Young women, aged 16 to 24 years, experience the highest rates of relationship violence (5). Approximately one in five female high school students report being physically and/or sexually abused by a dating partner (5). "(Female) adolescents and young adults are four times more likely to be victims of sexual assault than women in all other age groups (11)." Adolescent girls who report abuse from dating partners are more likely to use alcohol, tobacco, cocaine, or other types of substances (12). Furthermore, they are more likely to engage in risky sexual behavior, get pregnant, or consider or attempt suicide (12).

Studies show a relationship between media violence and certain violent behaviors of adolescents. Research on video games, violent content in rap music, pornography content (13) and beer advertising showed a likely association with aggressive behaviors among at-risk adolescents. Furthermore, long term effects of violent media are greater for children than adults (14). Many studies also conclude that adolescents are more vulnerable to negative media messages and perceive them as reality (6). Various theories focus primarily on television as the dominant cultural story teller of the age and predict that "viewers who watch a great deal of sex on television are more likely to accept the view of sex most frequently depicted on TV" (6). However, despite the abundance of teen dating violence information and research, the direct link between media influences and teen dating violence is not completely understood or fully established.

Outcomes

Findings from the literature review represent the input section of the logic model. As a result of these inputs, potential outputs or measureable objectives were then identified. These measurable objectives describe what the target population can achieve as a result of the activities and include:

- identify at least one way that media violence can affect teen dating behaviors,
- give an example of emotional abuse and an example of physical abuse that frequently occurs in dating relationships,
- state that teens can reject media messages that are harmful to healthy teen dating relationships,
- list the important rights that teens have in a dating relationship, and
- name one way to handle feelings that could lead to abuse or violence in a teen dating relationship.

Short-term goals or objectives were developed to facilitate the main purpose of the project. The expected outcomes of the activities include: 1) to provide information on teen dating violence, 2) to increase teens awareness of sexual and violent messages on male-to-female relationships in the media, 3) to help teens understand that violence in dating relationships can be influenced by media messages, 4) to discuss the causes of dating violence and the consequences of violence in a relationship, 5) to help teens explore alternatives to being abusive or violent, and 6) to establish a teen dating 'bill of rights'.

The long-term goals of the intervention are to include dating violence and abuse in the Media Madness curriculum for state wide distribution. As a result young people should be aware of the media influence on teen dating violence and to understand their options. In conclusion, the broad purpose of this project is to help break the cycle of disadvantage under the topic of self-worth.

Discussion

The literature manual was assembled for the practical application of the project and to also assist the staff at the Jane Fonda Center in future projects concerning teen dating violence. Many of the resources are excellent teaching tools for adolescents and will help guide various interventions aimed to improve their health. Listed in the manual are good models for building and improving healthy relationships between young people.

One aspect of this project was to explore the link between teen dating violence and low self-worth or self-esteem within the context of the CDD model. CDC states, "studies show that people who abuse their dating partner are more depressed, have lower self-esteem, and are more aggressive than peers" (7). Within CDC's Understanding Teen dating Violence Fact Sheet 2006, they define emotional abuse as threatening a teen or harming his or her sense of self-worth. They explain that the anger and stress that victims feel may lead to eating disorders and depression (15). Other "red flags" for dating abuse listed include: inability to manage anger or frustration, poor social skills, problems at school, lack of parental supervision or support, and witnessing abuse at home.

Implications from the teen dating violence literature review reveal that teen dating violence impacts females more than males. This raises questions for male perpetrators who could in the future continue to commit acts of violence. The research suggests that female youth suffer more from relationship violence than males. This fact poses essential challenges for adolescent intervention trials. The contention that low self-worth can be a risk factor for partner abuse among teens is a result of this project. Additionally, the fact that females are impacted by teen dating violence further reinforces the need to intervene before the pregnancy phase in the CDD. This information should help to narrow the focus of interventions on increasing awareness of youth, both males and females, of the CDD model and its correspondent adverse effects on adolescent health outcomes.

The success of the intervention depends on: 1) the assessment of the literature review or research on teen dating violence and any associations with the mass media, 2) the implementation of the activities in the Media Madness format, and 3) an evaluation of the outcome of the measurable objectives, short term goals, and four standards of effectiveness. The Media Madness format ensures that the section contains a manual, along with instructions for the leader or instructor.

Conclusion

Upon implementation of the intervention, we recommend a project evaluation, in order to determine if the activities were age-appropriate for middle school students, if the students were actively engaged, and if the activities are effective in increasing knowledge and promoting healthy attitudes. The measureable objectives and conceptual model of the intervention will also assist in planning the evaluation phase of the project. Overall, this project may help to engage, educate, and empower youth to break the cycle of violent unhealthy relationships by learning to resist unhealthy media messages. The next steps in this project will focus on how best to test this intervention and evaluate its impact. Based on the components of the highlighted project, this is a step in the right direction.

Acknowledgements

This project would not have been possible without the guidance, help, mentorship, and leadership of the staff at the Jane Fonda Center at Emory University. Special thanks to Marion Howard, PhD, Daniel Blumenthal, MD, MPH, and Jackie Davis, MPH, the Institute for the Study of Disadvantage and Disability and Morehouse School of Medicine's MPH Program, for their assistance with this project.

References

[1] Low self-esteem in teenagers. Professor's House 2007. Accessed 2013 Jun5. URL: http://www.professorshouse.com/family/children/lowself-esteem.htm
[2] Break the cycle II. Atlanta, GA: Inst Stud Disadvan Disabil, 2005.
[3] Brandy S, Matthew K. Effects of media violence on health related outcomes among young men. Arch Pediatr Adolesc Med 2006;160:341-7.
[4] Olden K, White S. Health related disparities of environmental factors. Med Clin North Am 2005;89:721-38.
[5] Rennison C, Welchans S. BJS special report: Intimate partner violence. Washington, DC: Bureau Justice Stat, 2000.
[6] Gerber G, Cross L. Living with television: the violence profile. J Commun 1976;1:172-99.
[7] CDC. Understanding teen dating abuse. NCIPC 2006. Accessed 2013 Jun 09. URL: http://www.cdc.gov/injury.htm
[8] NCVC. Dating violence. 2006. Accessed 2013 Jun 15. URL: http://www.ncvc.org/tvp.htm
[9] CDC. Youth risk behavior surveillance. Div of Adolesc Sch Hea 2002. Accessed 2013 Jun 05. URL: http://www.cdc.gov/HealthyYouth/yrbs.htm
[10] NCADV. Teen dating violence resource manual. Denver, CO: Natl Coalit Against Domestic Violence, 1997.
[11] Vicary JR, Klingamen LR, Hackness WL. Risk factors associated with date rape and sexual assault of adolescent girls. J Adolesc 1995;18:289-306.
[12] Silverman JG, Raj A, Mucci LA, Hathaway JE. Dating violence against adolescent girls and associated substance use, unhealthy weight control, sexual risk behavior, pregnancy, and suicidality. JAMA 2001;286:572-9.

[13] Brown JD, L'Engle KL, Pardun CJ, Guo G, Kenneavy K, Jackson C. Sexy media matter: exposure to sexual content in music, movies, television, and magazines predicts black and white adolescents' sexual behavior. Am Acad Pediatr 2006;117:1018-1027.
[14] Bushman BJ, Husemann RL. Short-term and long-term effects of violent media on aggression in children and adults. Arch Pedetr Adolesc Med 2006;160:348-52.
[15] CDC. Injury research agenda. NCIPC 2007. Accessed 2013 Jan 10. URL: http://www.cdc.gov/ncipc/pub-res/research_agenda/index.htm

In: Environmental Health
Editors: IL Rubin and J Merrick

ISBN: 978-1-62948-155-5

Chapter XII

Implementation and evaluation of a media intervention combating mass media's influence on the cycle of disadvantage and disability

Rasaan Jones, MPH*
Morehouse School of Medicine, Atlanta, Georgia, US

Research studies have explored the association between exposure to media violence and aggressive behaviors of youth, including relationship abuse. This project was an enhancement of a media curriculum that focuses on the subject of teen dating violence (TDV)/ intimate partner abuse (IPA), self-worth and the cycle of disadvantage (CDD) model. This chapter documents implementation and evaluation. The educational program (intervention) was pilot tested to evaluate feasibility and impact among the target group using two methods of program evaluation: process and impact. Methods: Twenty-one high school students in metropolitan Atlanta participated in a pilot test of the media intervention. The intervention group received the enhanced media curriculum over a two day period. At the end of the intervention, 8 of the 21 participants

* Correspondence: Rasaan Jones,MPH,Centers for Disease Control and Prevention, 1600 Clifton Road NE, Atlanta, GA 303291193 Chris Lane SW, Mableton, GA 30126 United States. E-mail: imn1@cdc.gov.

were selected to be a part of a focus group to measure if the evaluation questions were addressed and if the various objectives had been achieved. Results: The use of TDV/IPA case studies in the media was a useful approach for this intervention. The focus group identified emotional and verbal abuse as potential causes of physical abuse in a teen dating relationship. Impact evaluation suggests an increase in perceived risk of being in a TDV/IPA relationship among the focus group participants. However, some teens were already aware of media influences on aggressive behaviors and could identify media messages. Conclusions: The intervention highlights some positive outputs for project feasibility and impact. A pilot test evaluation was necessary but limited the assessment of reliability of the results.

Introduction

Researchers have identified two interrelated demographic trends with regard to adolescent behaviors: 1) increasing rates of youth violence and 2) increasing exposure of youth to mass media's violent content (1). It can be argued that the media reflects certain aspects of society as it really is, however through the process of observational learning and cultivation (1-2), certain aggressive behaviors can be negatively influenced by mass media. In addition, various attitudes, beliefs, and scripts presented through media violence promote and stimulate aggressive behavior (1). Children living in violent environments are at more risk to be influenced by media violence due to societal level risks (1).

Adolescents may benefit from interventions focusing on media messages and partner abuse because scientific evidence links exposure to media violence with aggressive youth behavior, including Intimate Partner Abuse (IPA) and Teen dating Violence (TDV) (1).When aggressive behaviors influenced by media exposure are not countered by a "norm against such behavior" it can increase the additive risk of aggression in some children (1). However, developing interventions to study and prevent at-risk behaviors among youth remain challenging in the initial development, implementation and evaluation phases.

The critical premise behind the Media Reflection Project states: "engaging, educating, and empowering youth to break the cycle of violent unhealthy relationships will assist in breaking the cycle of disadvantage and disability" (3). Several measurable, short-term objectives of this project developed during the planning stages of the intervention were evaluated using effective program evaluation methods. This was done to determine if the media intervention achieved the desired and intended effect.

The Jane Fonda Center at Emory's Rollins School of Medicine conceptualized and designed a media intervention model to provide an educational program that reduces the influence of the media on the sexual attitudes and behaviors of youth. The media reflection project was initially developed at the Jane Fonda Center and was an extension of the media madness curriculum. The curriculum was implemented by the Center with the intention of reaching 25,000 middle school youth throughout Georgia by training health educators, community stakeholders, and youth leaders to implement the curriculum. A third party evaluator is currently evaluating the curriculum.

The Media Reflection project is an enhancement to this model focusing on teen dating violence, media messages, adolescent self-worth, and the cycle of disadvantage and disability (3). The project is designed to assist youth in resisting unhealthy messages in the media and unhealthy, abusive relationships when dating. By increasing their perceptions and knowledge of TDV and IPA; identifying messages in the media pertaining to violent aggressive behaviors; and exploring alternatives to being abusive or violent in a relationship, destructive cycles of disadvantage can be broken or obstructed (3).

The overall project has two phases of development. The first phase of project development for the Media Reflection project details risk assessment, the media madness conceptual framework, and a literature review on media messages, TDV and IPA (3). Phase one of the project was published in the International Journal of Child and Adolescent Health in 2009 (3). It documents the initial development and planning of three activities using Jane Fonda's Media Madness / Getting the Rap on the Media model and conceptual framework.

Program planning and evaluation are important components for developing public health interventions. They both complement each other very well and comprise a vital phase of intervention development. Planning and evaluation help to ensure that public health interventions are logical, valid, and have measurable objectives or outcomes. Program planning should establish measurable objectives that are subject to evaluation. A thorough program evaluation will measure the effectiveness of an intervention designed to produce positive results among a target audience. Short and long-term goals and objectives are also evaluated to determine the efficacy of public health interventions.

Phase one concludes, “We recommend a project evaluation in order to determine if the activities are age appropriate for the audience, if the students are actively engaged, and if the activities of the media intervention are effective in increasing knowledge and promoting healthy attitudes”(3). Through various evaluation methods we can assess if specific guidelines of the intervention were followed, assess feasibility, and explore any short term impact.

Phase two of the project is the actual implementation of the activities (intervention) followed by a project evaluation. Documented in this manuscript, phase two describes the methodology used for pilot testing the three activities to a selected group of high school students in Atlanta, Georgia. Project evaluations are key to determining whether public health interventions are successful, therefore it was determined that the success of the intervention depended on the following characteristics: 1) feasibility 2) implementation of the activities in the Media Madness / Getting the Rap on the Media format and 3) feedback from the adolescents themselves.

In order to pilot test the intervention, two types of evaluations are selected for this study: process evaluation and impact evaluation. A process evaluation will determine if the intervention can be implemented as designed and is the most important form of evaluation for this stage of development. Feedback from the intervention group is deemed necessary to assessing if any of the major guidelines of the intervention are followed and to what level of satisfaction. Feedback is also important to assess any impact that the intervention may have on the target group.

This project evaluation should focus on addressing the following evaluation questions in regard to the media reflection intervention:

- Can a short intervention, lasting an hour and a half empower youth to identify media messages that can be harmful in terms TDV and IPA?
- How feasible is the intervention? Can it be implemented as designed?
- Do teens perceive themselves to be influenced by media messages?
- Do teens perceive themselves to be at-risk of being in an abusive relationship?

A media reflection: Phase one

The following sections recount vital information and components used for the initial curriculum development documented in phase one of the Media Reflection project. The logic model (see figure 1) and Media Madness formats guided the development of an innovative way to educate youth. This model also aided in developing the project outline, design, and construct variables that express the specific short and long-term project goals (3). The intervention curriculum guide provides a systematic process and instructions on how to conduct the activities.

Research on media influences and adolescent behaviors was retrieved from journal articles and databases during phase one of development (3). Identifying risks and protective factors was an essential component in the framework. A Medline search used the keywords: teen dating violence, teen dating abuse, teen relationships and violence and media and teen dating violence (3). This primary literature search enabled us to identify factors and behaviors associated with relationship abuse.

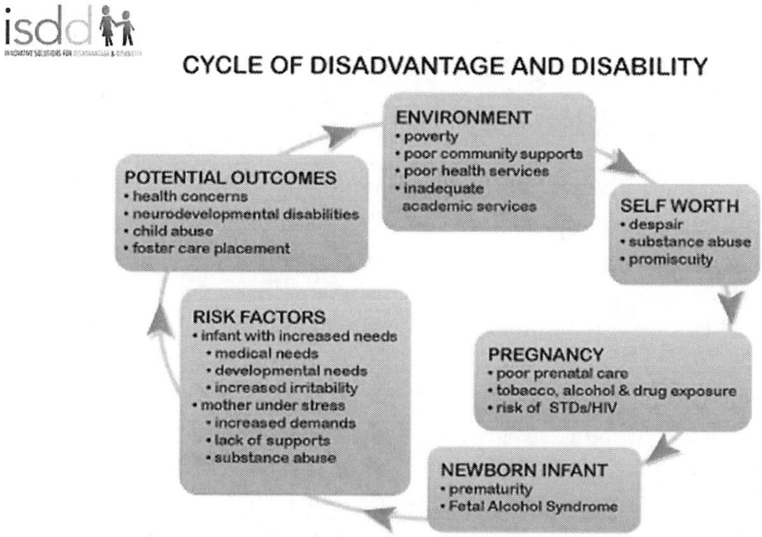

Figure 1. Cycle of Disadvantage and Disability. Break the cycle II. Institute for the Study of Disadvantage and Disability, 2005.

Four major guidelines were used as criteria for developing this intervention. First, there had to be an interactive portion of the activity; second, the intervention had to be fun and able to keep the youths' attention; third, the activities had to be age-sensitive regarding the subject material; and fourth there had to be a section for youth to express their opinions.

The first activity is a 15-minute lecture introducing the topic of teen dating violence to the audience followed by a "Myths and Facts" worksheet. The inputs for this activity come directly from the findings of the first literature review (3). The next activity includes an "agree or disagree" theme, which is an interactive activity using discussion topics about teen dating violence. This activity examines how youth need to think critically about common perceptions of abuse and violence within the adolescent community. This activity is supposed to encourage the participants to interact by having them move around and engage in a discussion. Each topic is posed as a statement to which they can agree or disagree with what is being stated. There is also a not sure option. Lastly is the creation of a teen dating bill of rights, which provides an empowering aspect of the intervention.

Institute for the study of disadvantage and disability

Public health organizations play a vital role in exploring the relationship between social and economic factors that affect children who are generally disadvantaged. The mission of the Institute for the Study of Disadvantage and Disability (ISDD) and its model is based on findings from an assessment of a clinic started at Hughes Spalding Children's Hospital in downtown Atlanta for children with cerebral palsy. The Institute reported that, "the findings [assessment of clinic] forced us to look more deeply into the relationship between social and economic disadvantage and the prevalence and severity of childhood disability" (4).

The goal of ISDD's Break the Cycle II project was to "raise awareness among future community professionals and academic leaders through encouraging the development of innovative and creative ways of overcoming barriers and developing services and supports for children and families at risk in their communities"(4). ISDD noted that children who are born into and who grow up in circumstances of social and economic disadvantage are more likely to have low self-worth, and this may lead to behaviors that perpetuate the cycle of disadvantage and disability (see figure 1) (4). The Media Reflection project was initiated under this proposal. It was developed using some of the constructs of ISDD's model of poverty and the resulting health implications.

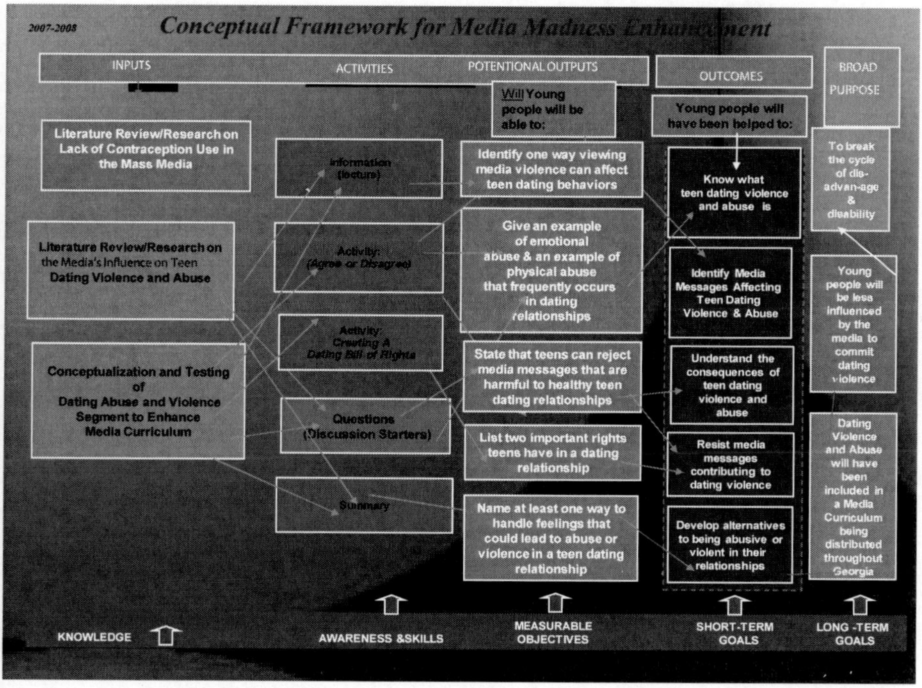

Figure 2. Conceptual Framework / Logic Model.

The CDD model and self worth

Figure 2 depicts the Cycle of Disadvantage and Disability (CDD). Within the CDD, the environment phase can pose many threats to the welfare and health of a community (4). The environment phase refers to issues of poverty, low socioeconomic status, poor health services, and community supports. The negative influences and behaviors shaped by these environmental factors are sometimes reflected in an adolescent's own self-worth, the next phase within the CDD model. The self-worth phase is important and critical to child development, as low self-esteem often elevates risk and exposure to substance abuse and can lead to pregnancy, which is the next phase in the cycle (3).

Low self-worth, despair, substance abuse, and promiscuity are all unhealthy feelings and behaviors that contribute to risky sexual behaviors and pregnancy. Within the CDD, low self-worth is a central risk factor for risky behaviors. Phase one of a Media Reflection poses the question: Is low self-worth a result of how people feel about themselves or a reflection of how others see them. Gerber's Cultivation Theory argues that viewers who watch a great deal of television are more likely to imitate what they see and perceive as reality (1). Planning for the Media Reflection intervention partly explored this bridge between CDD, mass media, and adolescent development using this theory of cultivation (3).

Literature review

Studies performed by Rowell Huesmann have demonstrated a relationship between media violence and certain violent behaviors of adolescents (1). Research on youth exposure to video games, violent content in rap music, pornography content, and beer advertising shows an association with aggressive behaviors (5). Furthermore, long-term effects of violent media are greater for children than adults (6) (7). The process by which children may become less sensitive to the pain and suffering of others is calleddesensitization and may play a key role in long-term effects of violent media on children(1).

Various theories focus primarily on television as the dominant cultural story teller of the age and predict that "viewers who watch a great deal of sex on television are more likely to accept the view of sex most frequently depicted on TV" (2). As a result, of this dominant influence, adolescents are more vulnerable to negative media messages and perceive them as reality (1) (2). However, despite the abundance of teen dating violence information and resources, researchers should focus on the link between media influences and teen dating violence (3).

Adolescents use music in the process of forming an identity (8). Data from the Council on Communication and Media support the notion that the prevalence of music listening among adolescents is higher than television viewing (9). They have concluded that exposure to violence, sexual messages, and sexual stereotypes in music videos produce significant changes in behaviors and attitudes in adolescents (9). In addition, music lyrics have also become more explicit in their references to sex and violence in certain genres (9). In 1999, The National Institute on Media conducted content analysis of the top ten music CDs, and found that each CD included at least one song with sexual content. They revealed that rap music has been characterized with messages of violence and hatred toward women (10). Of 518 music videos sampled in their study, the percentage of violence in the music videos ranged from 11.5% to 22.4%. It was determined that rap music videos had the highest portrayal of violence compared to all other genres (22.4%) (10).

Teen dating violence and abuse

Dating violence is defined as controlling, abusive, and aggressive behavior in a romantic relationship. It can include verbal, emotional, physical, sexual abuse, or a combination of all. The National Center for Victims of Crime explains that, "Both boys and girls can be victims of dating violence, but boys and girls abuse their partners in different ways" (11). They state, "Boys physically injure girls more, thus are more likely to punch their partners, and boys also force their partners to participate in unwanted sexual activity, while, "Girls are more likely to yell, threaten to hurt themselves, pinch, slap, scratch, or kick their partners" (11).

Some teen victims experience violence occasionally, while others are abused more often, sometimes daily. However, "one in five teens in a serious relationship report having been hit, slapped, or pushed by a partner" (11). Using a national sample, the CDC (2000) reported that the average prevalence of dating violence for high school and college students is 22% and 32%, respectively (12). As a dating relationship becomes more serious, the potential for and nature of violent behavior escalates (13). Approximately one in five female high school students report being physically and/or sexually abused by a dating partner (14).

More importantly, abuse from dating partners can significantly elevate risk for a board range of serious public health concerns and burdens (13). For example, young women, ages 16 to 24 years, experience the highest rates of relationship violence (14). "(Female) Adolescents and young adults are four times more likely to be victims of sexual assault than women in all other age groups" (15). Adolescent girls who report being abused by their dating partners are more likely to use alcohol, tobacco, cocaine, or other types of substances. Furthermore, they are more likely to engage in risky sexual behavior, get pregnant, or consider or attempt suicide (16).

Vulnerable populations

A study by the Kaiser Family Foundation reported that youth spend an average of 6 ½ hours a day with some type of media and the majority of youth say their families have no rules about their media use (17). African American youth spend the most time watching TV, music videos, DVDs and movies. The study also revealed that African American kids are more likely than White kids to report having televisions, DVDs, cable/satellite TV connections, subscriptions to premium TV channels, and video game consoles in their bedroom (17).

June Price Tangney and Seymour Feshbach reported that Black youth "watch nearly twice as much TV as other children, regardless of parents' education, sex, age, and family composition". They suggest one reason may be that there's a greater acceptability of television as a source of information and entertainment in the Black community (18).

Young people in the lowest income group are more likely to have their own TV in the bedroom compared to middle or higher income levels (17). The Kaiser Family Foundation concluded that increased media consumption decreased activity levels in youth, increased exposure to unhealthy food consumption, impacted alcohol use and sexual activity among African American youth, and influence girls perception of body image (17).

Media exposure and relationship abuse among students

The Youth Risk Behavior query tool available through Online Analytical Statistical Information System (OASIS) within the Georgia Division of Public Health reports that in 2007, 61.6% of African American Georgia middle school students watched three or more hours of TV on an average school day as compared to 28.5% of white students. Sixty-four percent of African American high school students watched three or more hours of TV per day on an average school day in Georgia compared to 28.6% of White students (19).

The percentage of high school students who reported being hit, slapped, or physically hurt on purpose by their boyfriend or girlfriend during the past 12 months in Georgia for the year 2007 was 18.6% for African Americans compared to 12.7% for Whites. For African American males, the percentage is 19.5% and for African American females, the percentage is 17.6%. The percentage for White males is 12.7% and 14.0% for White females (19). The percentage of African American middle school females students who had reported ever been in a physical fight in 2007 was 76.3% and 87.2% for males. While the percentage of white students for the same category is 54.1% for females and 69.8% for males.

All statistics for TV media exposure on an average school day and physical violence among peers are higher for African American students in Georgia for 2007. No direct correlation between media exposure and violence or relationship abuse among students in Georgia has been reported. However, a relationship abuse curriculum focusing on media exposure and aggression may directly benefit African American students within the state of Georgia.

Interventions

Desensitization and Gerber's cultivation theory are the processes by which youth's behavior may be affected by exposure to media. In addition, the long-term effects of violent media on children can lead to social conflicts (6). Adolescents and young people are vulnerable to violent media exposure. More focus and effort should be put on addressing the link between TDV/IPA and media messages.

Abuse from dating partners can significantly elevate risk for a board range of serious public health concerns. Adolescent girls who report being abused by their dating partners are more likely to use alcohol, tobacco, cocaine, or other types of substances (15). The fact that females can significantly be impacted by health implications and burdens of teen dating violence further reinforces the need to intervene before the pregnancy phase in the CDD model.

The relationship between media violence and aggressive behaviors of youth can profoundly impact African American youth (18). Data from high school students in Georgia reveal that media exposure in the form of TV viewing on an average school day is higher for Blacks than Whites. In addition, more African Americans reported violence from a dating partner than other races in 2007 (19).

Teaching youth about healthy relationships as part of their required health curriculum can reduce Partner Dating Violence and increase condom use 2.5 years later (20). The inclusion of "additional training in the dynamics of dating violence and healthy relationships" can decrease teen dating violence among adolescents (20). Information on media violence should help to narrow the focus of interventions to increase awareness on media messages and media literacy. As previous noted, these innovative media interventions have to be tested and evaluated as part of the overall program planning process. Phase two of a Media Reflection reports the feedback given from African American high school students in Atlanta. An evaluation of this media intervention should highlight any short-term impact on the students.

Phase two study

The current study (Phase 2) utilized qualitative methods to pilot test implementation of the Media Reflection project, which had not been previously implemented among African American teens. IRB approval was granted from Morehouse School of Medicine to conduct this evaluation. A convenience sample of twenty-three African American high school students in metropolitan Atlanta agreed to participate in the study. All participants in this study were recruited from a band class representing a local high school in Atlanta.

Procedures

The intervention was conducted over two days. On the first day, participants participated in the lecture on TDV/IPA myths and facts, and completed the teen dating bill of rights exercise. During day two, participants engaged in the "agree or disagree" TDV/IPA discussion and debate. Each student was instructed to indicate their position following the reading of various statements on TDV/IPA by standing next to agree, disagree, or not sure signs.

Both evaluations were facilitated by conducting a focus group with the participants at the end of the intervention. Eight participants were selected to participate in the focus group at the end of the intervention. The focus group was composed of 4 males and 4 females. Feedback from high school students was a key element of the process evaluation and pilot test of the media intervention. A focus group guide was used by the moderator and co-moderator to provide a systematic and structured qualitative inquiry (see figure 3).

Qualitative measures

Focus group questions were separated into three areas of post intervention assessments. The first set of questions related to assessing the participants' perceived risk of being in a TDV/IPA relationship. The second set asks about the guidelines of the intervention. The last set is composed of questions designed to help assess the students' perceptions of media influences, messages and aggressive behaviors in a dating relationship. All focus group questions correlate to the major evaluation questions of phase two.

Set 1: Relationships, TDV/IPA and perceived risk

Key constructs of the Health Belief Model were used for this section. Questions regarding perceived risk of being in abusive relationship were collected by asking the students' opinions, experience and attitudes related to TDV/IPA and abusive relationships among high school students or peers. Students who participated in the focus group were first asked to explain how they felt about a recent high profile case of relationship abuse in the media involving music singers Chris Brown and Rihanna. Due to the popularity and dynamics of this case, many teens were familiar with the incident. This question was designed in the focus group guide as a key, introductory, topic question which would get their attention and spark interest. Other questions within this domain were related to assessing opinions on TDV /IPA, teen relationships and perceived risk. Participants were asked if TDV/IPA was a problem almost teens, peers, and people they know. Questions in this set also asked if teens in their environment or school date a lot and have a lot of relationships.

Set 2: Intervention guidelines, referrals and improvement

To help assess the feasibility and impact of the intervention, participants were directly asked if any of the activities were useful to them. In addition, they were asked if the activities followed guidelines or espoused characteristics of activities including whether they were fun, interactive, age appropriate, and to rate the activities using a scale of 1 to 5, 1 being most helpful and 5 being least helpful. The second key question in this set asked the participants if they would share any of the information from the activities with friends who were in an abusive relationship. In conclusion, the last question asked the participants how they would improve the activities.

Set 3: Media exposure, media messages and aggressive behaviors

The last set of questions correspond to the major evaluation questions related to whether youths can identify media messages that can influence aggressive behaviors in a dating relationship. Direct responses were recorded as to whether any of the activities could help them identify these messages or behaviors in the media. Participants were then asked if they think someone can become violent or aggressive by watching violence on TV or by listening to violent content in music.

All responses were recorded by an audio device to provide an accurate and permanent record of the discussion. A moderator facilitated the focus group discussion while a co-moderator took notes to capture other verbal and non-verbal communication in response to the questions asked.

Data analysis

Of the 23 students who agreed, 21 participated in the intervention. The majority of participants (56.3%) were in 9th grade. All data, preliminary results and feedback relate only to the intervention group to evaluate feasibility. The focus group transcript was developed as a basis for data analysis. The research team developed a codebook based on the focus group questions and major evaluation questions where words and themes of the participants' responses, through verbatim transcript responses, were categorized. These words and themes were later analyzed and coded. Major themes were identified and synthesized using standard content analysis procedures. For example, words and phrases that were unfamiliar or used in a new way were recorded on the codebook. Interpretation of the phrases or results was based on the range of variation among the focus group members and identified patterns in the codebook. The level of agreement among the focus group members and corresponding questions were noted.

A Media Reflection: Implementation and Evaluation Focus Group Guidelines & Questions
Note: Remember to remind teens that their feedback is very important and it is okay to speak their minds. The purpose of this focus group is to get their opinions about the following topic below. Encourage teen to speak their minds. Remind participants that the focus group is voluntary. Each group should be diverse representing both boys and girls. The focus group is designed to last 30- 45 minutes.
Read: The reason for this group discussion is to get your opinions, ideas, and feelings about the activities we have been doing in this class. I also want to hear what you think about teen relationships, the media, and violence in music and TV. There are no right or wrong answers!
Introduction Questions: Relationships, Teen Dating Violence, at risk
(Key) Chris Brown and Rihanna (case studies/ get opinions) –
1) What do you guys think about the Chris Brown and Rihanna case last year? (Probe) Is this a case of dating violence? Who was right or wrong? Why?
2) What do you think about teen relationship abuse? Is abuse or violence a problem for teens in a relationship? Do teens see and copy these kinds of behaviors?
(Key) Important rights for dating
3) Do teens in your school and/or community date and have relationships a lot? Do you think teens need rules and rights in a dating relationship? Why? What are some rights that teens have when going on dates and being in relationships?
(Key) Example of emotional abuse & physical abuse -
4) What are the differences between emotional abuse and physical abuse? Which type is worse? Do girls tend to emotionally abuse boys more? Do guys use physical abuse more?
Phase 1 Questions: Activities / Guidelines - feedback
(Key) 3 activities - interactive, fun, attention, age sensitive, opinions
5) Were any of the activities useful for you? Which ones? What way? If not, why? What did you get out of the activities (probe with follow-up questions, ask for more input). Were any of the activities fun? How would you rate the activities? - between 1 and 5, 1 being real boring and 5 being very exciting. (Probe) Did you feel like the activities got your attention? Why? Were the activities okay for your age group (9th graders)?
6) Is there anything new you learned? What did you know about TDV before the activities?
(Key) refer friends
7) If you saw a friend being shoved and slapped by a boyfriend or girlfriend, would you tell them about this group of activities? Or what you learned? Why? Why not?
(Key) how to improve the activities
8) If you could improve these activities, how would you do it? What would you add?
Phase 2: Media Exposure / Media Messages / Aggression Behaviors
(Key) Identify a way viewing media violence can affect dating behaviors
9) Do you think teens are influenced by the media (TV music)? What about bad things?
10) Do you really think someone can become violent or aggressive by watching violence on TV or listening to violent music.
11) Did any of the activities help you identify these messages or behaviors in the media?
12) Do you think TV shows life as it really is?
13) Do you think teens need activities and education to help them with dating? Do you think teens need more information about how the media can influence them to do risky things?

Figure 3. Focus Group Guide.

Findings

The results of the process and impact evaluation are presented in qualitative form as quotes extracted from the audio recordings of the focus group (see table 1, 2, & 3). The displayed quotes belong to members of the focus group in response to key questions from the focus group guide (figure 3). Tables 1, 2, and 3 display the questions and feedback in the codebook. Similar responses, repeated responses, and certain responses to follow-up probes were not included into the table. Three key themes correspond to three different sets of questions asked on the guide in order to address the evaluation questions of this project.

Table 1. Focus Group Student Responses- Set 1: Relationships, TDV/IPA, and Perceived Risk

1. Case study: Chris Brown and Rihanna- What do you guys think about this case?
I think it kinda depends too…because Rihanna was kinda at fault too because she was verbally abusing him and so that caused him to hit her.
Kids joke around a lot with it…like…the boys say..like I'll Chris Brown you if you don't shut up.
Probe Question: Who was right and who was wrong?
I think it's also Chris Brown's fault because he has the obligation to be a man and say no, I'm not going to hit you.
I don't think Chris Brown is more at fault than Rihanna it had to take something to set him off like that whether its physical or verbal
Physical and verbal is almost the same thing
A girl can't consistently yell at a guy and think he won't retaliate.
2. Do you think TDV and IPA is a problem amongst teens, peers, and people you know?
I do, I feel like teens in relationships are more unstable than grownups because they don't have that level of maturity
They are really unstable because they're not mature enough. Some girls are really strong but other girls are easy broken…They're emotions are not strong enough.
I'm glad you said that [responding to the last comment] because a lot of teens emotions are unstable before they get in a relationship.
They [teens in relationships] end up more hurt than grownups can be.
3. Should there be rules and rights in teen relationships?
They shouldn't be a relationship if they can't handle a relationship.

As a teenager, we don't know how to maintain a good relationship...we don't know what it takes to keep a relationship.
I really think there should not be any rules in a relationship cause, I mean, if a girl set a rule on me and I can't like give her a hug or put my arm around her I feel like I'm hinda getting neglected.
Probe Question: Which is worst emotional abuse or physical abuse?
I think emotional abuse is worst because with physical...that can always hurt but they will heal over time but you can be emotionally scarred for life.
The difference is like... when you talking about her or name calling..like that could stay in her head forever.
You can always recover from physical abuse.

Table 2. Focus Group Student Responses- Set 2: Media Reflection Activities, Referrals & Improvement

1. Were any of the activities useful for you?
I think it was good because teens should know about this stuff and discuss it.
No, I don't think it was useful for me because I m not in a relationship.
I think it was useful because you change somebody's life that's going through abuse.
I think this was really useful, you engaged and I was not bored... it was pretty good...I learned a lot...the noise was a problem.
I would rate it a 4 1/2, the half is because of the noise.
I think it was really good...because it opens your mind to what's going on in the world and stuff so on a scale from 1 to 5, I would rate it a 4 1/2
I think it was useful because you always need reinforcement on a subject like this because it can happen to you, it can happen to anybody you know and it can help you out
I would say it's a good preview...for study on too it [more study on the subject]
2. Refer friends- If you saw someone you knew abusing their partner would you tell them about this group of activities?
Unanimous agreement : *Yeah*
3. I would say yeah because...obvious they're doing it and most of the time they don't know they're doing it, so it takes someone to tell them cause how do you know when you're wrong when you don't acknowledge it yourself.
What's something we could improve regarding the activities? Is there something we could have done better?
You could like bring in somebody...a boy and a girl who has been in a physical and emotional relationship...and get their point of view.

Table 3. Focus Group Student Responses- Set 3: Media Exposure & Aggressive Behaviors

1. Do you think teens are influenced by the media?
I say yes because some people just watch TV and do whatever it tells them to do...Say like Little Wayne, you see him sagging and people going to sag too ...you see him talking about other stuff, they probably do the same thing too... I've heard people trying to... be like him.
I still say no because either way, you may think it's fun but you still have that second thought, that say maybe this don't work or this doesn't happen like that...it's still they're fault and either way you can't blame it on the media, good or bad.
2. Do you think someone can become violent or aggressive by watching violence on TV?
I do, because its someone that you look up to, like a role model and you see them beating on a girl...and you think...its cool for them so what wrong if I do it...
Probe Question: Do you think the activities will help you identify these messages or behaviors in the media
Unanimous agreement: Yeah
3. Does TV show life as it really is?
Unanimous agreement: No
What people want to see, what they wish life was like...reality TV, I think that's fake because nobodies life is like that...TV s is like a big imaginary world filled with things you might like, they might show like the perfect family.
TV is just trying to get ratings

Participants' responses to the Chris Brown and Rihanna questions hinted that verbal abuse may have occurred between the two of them before any physical abuse. Verbal abuse was a major source of interest for the students and they hinted that this may cause teens to react with physical abuse. When asked the follow-up question, "who was to blame", most agreed that both parties shared the blame. One participant stated that, "physical and verbal abuse was almost the same thing". However, there was some variation in regards to the seriousness of verbal or emotional abuse versus physical abuse.

All responses indicated that teen relationships can be abusive and that teens are susceptible to being in abusive relationships due to their instability and immaturity. Participants believed that teen relationships "are really unstable because they're [teens] are not mature enough. Some girls are really strong". Others explained, "I do, I feel like teens in relationships are more unstable than grownups because they don't have that level of maturity".

As a follow up, participants were asked if teens needed rules and/or rights when dating. Many of the responses did indicate that teens do date very often and are in a lot of relationships. One participant explained, "As a teenager we don't know how to maintain a good relationship…we don't know what it takes to keep a relationship". However, when talking about teens having rules and rights in a dating relationship, one key responded mentioned, "I really think there should not be any rules in a relationship cause, I mean, if a girl set a rule on me and I can't like give her a hug or put my arm around her, I feel like I'm kinda getting neglected." The trend of responses agreed that some rules are needed in teen relationships. When asked who is more likely to be abused emotionally: males and females, all participants agreed that females were more likely to receive emotional abuse from a dating partner.

All but one response indicated that the activities were useful to them and potentially would be for their peers. The information given during the intervention was perceived to be relevant. For example, one participant said, "I think this was really useful, and engaging and I was not bored…it was pretty good…I learned a lot…the noise was a problem". Participants mentioned that there was too much noise during several activities and that was the only drawback. However, the noise was due to the strong participation and engagement of youth. Participants also wanted the intervention to be longer and were intrigued by the subject matter. As one quote elucidates, "I think it was really good…because it opens your mind to what's going on in the world…. so on a scale from 1 to 5, I would rate it a 4 ½".

All participants agreed that they would refer the activities or the information received from the intervention to a friend or peer who they know was in an abusive relationship. One response says, "I would say yeah because…obvious[ly] they're doing it and most of the time they don't know they're doing it, so it takes someone to tell them cause how do you know when you're wrong when you don't acknowledge it yourself." All but one of the participants did not recommend a way the intervention could be improved due to the lack of responses to this question.

There was some disagreement as to whether the media directly influences teens and if the media is somehow responsible for the actions of others. One response said that it does, "I say yes because some people just watch TV and do whatever it tells them to do…Say like Little Wayne, you see him sagging and people going to sag too …you see him talking about other stuff, they probably do the same thing too... I've heard people trying to… be like him." References to music stars were frequent when the students discussed media messages and influences. One conflicting response disagreed and mentioned that people cannot blame the media for their actions; however the consensus was that the media does influence teens.

When asked if the intervention or activities helped them identify messages and behaviors in the media related to TDV/IPA relationships, there was unanimous agreement that it did. Key terms and phrases like reality TV and references to music role models were recorded when the participants were asked if the media portrays real life as reality.

Discussion

All participants in the focus group agreed that the activities helped them identify messages in the media that could influence aggressive behaviors in a dating relationship. However, some of the participants explained that youth are responsible for their own actions and should be able to make decisions in spite of the media. This theme points to some level of self-dependence and confidence that teens can resist the media with media interventions. Many of the teens were already aware that the media can influence them in negative ways based on the responses and observations during the teen dating discussion activity (see figure 3). The students also responded unanimously that the media does not show life as it really is and mentioned the terms "entertain" and "entertainment" when describing the functions of the media. However, all agreed that the intervention was useful for them and that they would refer this information to others.

Various phrases recorded from the focus group indicate that the intervention was feasible and could be implemented as designed given certain time considerations. Additional evaluation on program feasibly should be conducted by implementing the intervention in a one day, one session format to compare with the two day format used during this pilot test. Not many recommendations for the intervention were offered by the focus group members and a major limitation of the study was that all intervention participants did not provide feedback or were in the focus group.

Another limitation of the evaluation is that the design is a pilot test; therefore more assessment is needed to fully determine its impact and feasibility. In addition, the sample size of high school students for this evaluation is far too low to establish reliability for the target population. The participants' feedback may not be representative of a lot of high school students in Atlanta. The results of the evaluation strictly relate to the intervention group and focus group participants.

One major outcome of the intervention was that the subject of teen dating violence was extremely relevant for this population. The introduction of the subject of dating and TDV/IPA got the students' attention by conducting the myths and facts exercise. This exercise encouraged the intervention group to be actively engaged, as evidenced by the high participation level of the students. The Chris Brown and Rihanna case is an important example of how the subject of TDV/IPA and the influence of the media are connected. The case was also very useful for getting the students' attention and feedback regarding TDV/IPA. Therefore, case studies may be a very effective teaching tool in media interventions and the media madness activities, as many of the intervention participants became actively engaged in this discussion. It is important to note that most of the participants did not know the full details of the incident. However, the common themes and terms surrounding their responses seem to focus on the notion that Chris Brown, being the perpetrator of violence, was not fully at fault and part of the blame could be placed on Rihanna.

The students' responses from this media case also highlight a very important outcome of this evaluation regarding verbal or emotional abuse and physical abuse. The logic model for the intervention indicated that the definitions of emotional and physical abuse are key outputs (1). Responses to this set of questions may suggest that the students feel that verbal abuse is a predictor of physical abuse. Terms like "verbal abuse", "depends", "blame", "fault", and "causes" were collected in the codebook. These terms were most common during this set of questions. It is interesting to note that many of the students thought that Chris Brown was somehow justified by responding with physical violence because they believe he had been verbally abused by Rihanna. The majority of responses seem to indicate that receiving verbal abuse from a partner may be an initiator or justification for responding with physical abuse.

Conclusion

Finally, based on the pilot test of the intervention, the Health Belief Model emerged as a useful theory which can improve the outputs of this project when used in conjunction with the logic model for development. This theory was not initially used in the development of phase one of the Media Reflection. However, key constructs such as perceived susceptibility and severity of being in a violent or abusive relationship were used for evaluating the impact of this intervention. This theory should be explored when developing interventions that focus on healthy relationships and used to break the cycle of disadvantage and disability. The cultivation theory still remains relevant, however many of the intervention participants did not think TV shows life as it really is. The intervention could benefit by adding activities that provide instruction on coping with aggression in a dating relationship. A satisfaction survey rating the intervention is highly recommended.

Further evaluation for long-term impact is needed to determine if the students will avoid bad relationships. The pilot test only evaluates real short-term objectives as related to the small sample size of the intervention group. This intervention may best be served as a reinforcement tool to increase perceptions of risk. It should also identify barriers to having healthy relationships among teens. Youth can avoid harmful TDV/IPA relationships by receiving some of the information and case studies in this intervention.

Acknowledgements

This project would not have been possible without the guidance, help, mentorship, and leadership of my Culminating Experience Committee, the staff at Morehouse School of Medicine's Master of Public Health program, and the staff at the Jane Fonda Center at Emory University. Special thanks to Tabia Akintobi, PhD, MPH, Elaine Archie Booker, EdD, RN, Paul Abamonte, MBA, MS, Barbara Moore, Marion Howard, PhD, Daniel Blumenthal, MD, MPH, and Jackie Davis, MPH, the Institute for the Study of Disadvantage and Disability, for their assistance with this project.

References

[1] Husemann LR, Moise JF, Podolski C. The effects of media violence on the development of antisocial behavior. New York: John Wiley, 1997.

[2] Gerber G, Cross L. Living with television: the violence profile. J Commun 1976;1:172-99.

[3] Jones R. A media reflection: Combating mass media's influence on the cycle of disadvantage and disability. Int J Child Adolesc Health 2009;2(3):285-92.

[4] Break the cycle II. Atlanta, GA: Inst Stud Disadvan Disabil, 2005.

[5] Brown JD, L'Engle KL, Pardun CJ, Guo G, Kenneavy K, Jackson C. Sexy media matter: exposure to sexual content in music, movies, television, and magazines

predicts black and white adolescents' sexual behavior. Am Acad Pediatr 2006;117:1018-27.

[6] Bushman BJ, Husemann RL. Short-term and long-term effects of violent media on aggression in children and adults. Arch Pedetr Adolesc Med 2006;160:348-52.

[7] Brandy S, Matthew K. Effects of media violence on health related outcomes among young men. Arch Pediatr Adolesc Med 2006;160:341-7.

[8] North A, Hargraves J, O'Neil A. The importance of music to adolescents. Br J Educ Psychol. 2000;70:255-72.

[9] Council on Communication and Media. Impact of music, music lyrics, and music videos on children and youth. Pediatrics2009;124:255-72.

[10] Durant H, Rone S, Rich M, Allred E, Woods R. Tobacco and alcohol use behaviors portrayed in music videos: A content analysis. Am J Public Health 1997;87:1131-5.

[11] NCVC. Dating violence. 2006. Accessed 2013Jul 15. URL: http://www.ncvc.org/tvp.htm

[12] CDC. Youth risk behavior surveillance. Div of Adolesc Sch Hea 2002. Accessed 2013Jul 05. URL:http://www.cdc.gov/HealthyYouth/yrbs.htm.

[13] NCADV. Teen dating violence resource manual. Denver, CO: Natl Coalit Against Domestic Violence, 1997.

[14] Rennison C, Welchans S. BJS special report: Intimate partner violence. Washington, DC: Bureau Justice Stat, 2000.

[15] Vicary JR, Klingamen LR, Hackness WL. Risk factors associated with date rape and sexual assault of adolescent girls. J Adolesc 1995;18:289-306.

[16] Silverman JG, Raj A, Mucci LA, Hathaway JE. Dating violence against adolescent girls and associated substance use, unhealthy weight control, sexual risk behavior, pregnancy, and suicidality. JAMA 2001;286:572-9.

[17] Kaiser Family Foundation. Generation M: Media in the lives of 8-18 year olds. Menlo Park, CA: Authors, 2005.

[18] Tangney JP, Seymour F. Children's television-viewing frequency: Individual differences and demographic correlates. Pers Soc Psychol Bull1988;14:147.

[19] Georgia Department of Human Resources, Division of Public Health, Office of Health Information and Policy. Online Analytical Statistical Information System.YRBS.Accessed 2013Mar 04. URL: http://health.state.ga.us/epi.cdiee/studenthealth.asp.

[20] Wolfe D, et. al. A school-based program to prevent adolescent dating violence. Arch Pediatr Adolesc Med 2009;163:692-9.

In: Environmental Health
Editors: IL Rubin and J Merrick

ISBN: 978-1-62948-155-5

Chapter XIII

Cyclopedia: Sustaining a positive youth development program

Cappy Collins, MD**, *Marc Lavender, MD, Shawn Brown, BA, Perry Sheffield, MD, MPH and C Andrew Aligne, MD, MPH
Department of Preventive Medicine, Mount Sinai School of Medicine, New York, Golisano Children's Hospital at Strong, University of Rochester, Rochester, Rochester Boys and Girls Club, Rochester, New York, US

Urban, low-income minority adolescents are disproportionately affected by numerous preventable morbidities, including poor school performance, obesity and teen pregnancy. Due to the complex interplay of social, cultural and economic factors that contribute to the morbidities, no single intervention has shown consistent success in categorically improving outcomes. Recent research suggests a common root problem is lack of empowerment. Effective positive youth development (PYD) initiatives can prevent many morbidities by fostering empowerment in at-risk youths. However, little is published on how to implement and sustain such interventions. Methods: Cyclopedia is an annual summer curriculum of mentored bicycle rides consisting of educational lessons, local

* Correspondence: Cappy Collins, MD, Fellow, Pediatric Environmental Health, Mount Sinai School of Medicine, Department of Preventive Medicine, 19 East 102nd Street, Floor 3 West, New York, NY 10029 United States. E-mail: cappy.collins@mssm.edu.

exploration, and creative self-expression (photography and writing). It is structured on a PYD model and is supported by a core partnership between a medical institution and a community-based organization (CBO). Results: Cyclopedia has completed two full seasons with expansion from 18 boys to 50 boys and girls. The cost per trip has dropped from roughly $400 per rider to $100 per rider. The number of rides increased from 13 to 20, and total mileage increased from 740 to 870 miles per year. However, miles per rider dropped from 41 to 17, and writings per rider dropped from 2.6 to 0.5. Conclusions: Cyclopedia, supported by a core partnership between a medical institution and a CBO, shows significant promise in cost-effectively reaching large numbers of at-risk children with PYD programming. Ongoing curriculum adjustments aim to address decrements in individual participation.

Introduction

Almost 40 years ago, Robert Haggerty coined the term "new morbidity" to describe the emergence of a new set of diseases afflicting children in the second half of the 20th century (1). Unlike vaccine-preventable and other acute infectious diseases, these new morbidities were typically chronic in nature, with strong behavioral and psychosocial components. Gazing into his "crystal ball" in 1974, Haggerty appears prescient: "We are going to be dealing in the future with...prevention of diseases that will not appear until the adult period of life—atherosclerosis, obesity, smoking, and lung cancer, diseases that will require different ways of living from childhood on." The paradigm shift away from "waiting until the child comes in to the office, and ignoring all the ones in the community who do not come," required a new strategy for delivering care. He endorsed "community pediatrics" as a team-based approach for providers to not treat just the acutely ill patient, but to treat the child, family and society chronically beset with health disparities.

Research in the subsequent decades suggests that the new morbidity is not equitably distributed. Chamberlain et al (2) summarized the situation:

> Across nearly all illnesses, poor health outcomes are closely associated with a declining gradient of family income (3-5). Independent of income status, children from ethnic minorities are more likely than non-minority children to suffer from asthma, heart disease, obesity, diabetes, injury, violence, poor preventive care, and a host of other significant health risks (6-7). Other related root causes of child morbidity

include limited access to care, low social capital, low caregiver literacy, and unsafe neighborhoods (8-11)

More recent research suggests urban, low-income minority populations are also disproportionately affected by poor school performance, violence and teen pregnancy (12-16).

Rubin et al. (17) have described these links between the physical, political, economic and social facets of a child's environment as the "cycle of environmental health disparities." Given the complex interplay of these factors, Haggerty's community-based, "team" approach, indeed, would seem an appropriate strategy for addressing the unjust burden of disease on residents of impoverished communities (see figure 1).

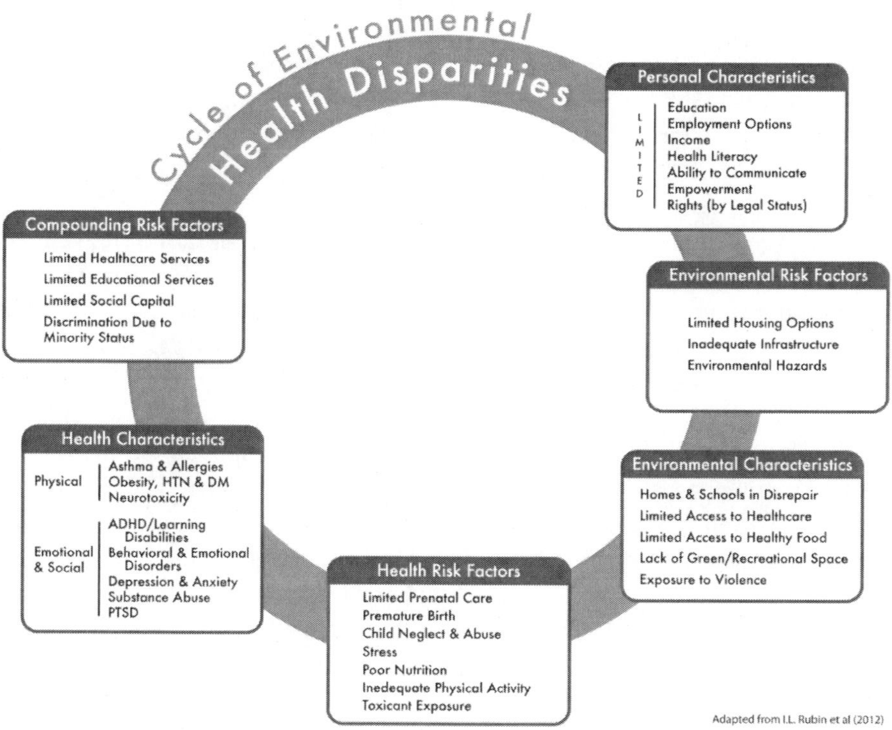

Figure 1. Cycle of Environmental Health Disparities.

In summary, the key to the health of poor, urban and minority children lies not simply in treating disease, but in protecting them against the effects of unjust conditions and, ultimately, in alleviating the injustices. In stronger terms, the described morbidities represent failures of our current healthcare efforts.

On the positive side, there have been success stories. As vulnerable as children are during their development, they can be protected. The Nurse-Family Partnership has shown that intervention during the prenatal and early childhood periods can benefit children even into adulthood with reducing criminal arrests and pregnancies (18). It is likely that the community-based approach of partnering with the young parents of these children is crucial to the program's success.

Adolescence essentially represents the last opportunity for pediatric providers to intervene. Developmentally, this is the time to support the efforts of youth to become productive and healthy adults who can provide for their own needs and contribute to their communities.

Due to the complexity and variety of problems, no single intervention has shown consistent success in categorically improving outcomes in at-risk adolescents. However, a growing body of research has suggested, first, that a shared root problem is lack of empowerment and, second, that effective positive youth development (PYD) initiatives are a promising model to improve outcomes by addressing that root problem.

Positive youth development is a broad term to describe preventive initiatives to help youths build on their personal assets, family support and community resources in order to maintain health and healthy behaviors as they transition from childhood to adulthood. There are various PYD frameworks endorsed by various research groups; the core components include support from positive adult role models, education, a safe and structured environment, and opportunities to link self-identity with the larger community.

PYD strategies are a diverse collection of interventions. A 2002 report from the National Research Council and Institute of Medicine documents dozens of PYD programs that have been effective in improving outcomes for adolescents: school attitudes, school attendance, school performance and school violence; attitudes towards adults, the elderly and community service; tobacco, alcohol and other substance use; family conflicts and aggression; self-control, confidence and self-esteem; communication skills; bullying and socialization; pregnancy and sexually risky behavior; mood disorders and risk for suicide (19).

The Boys and Girls Clubs of America, in particular, has shown success with Project Learn, a PYD initiative that enhanced academic performance in at-risk, minority (63% black, 19% Hispanic) youths living in public housing in

New York City, Cleveland, Oakland, Tampa and Edinburgh, Texas (20). Additionally, several state legislatures have endorsed PYD programming (21).

These myriad PYD initiatives are all very different in design. Each one was developed to accomplish particular objectives, and uses different tools to implement the PYD components. Characterization of what makes each program sustainable or successful is lacking, aside from simply qualifying as a PYD program.

Since there is no one template for successful PYD programming, we hypothesize that a PYD empowerment program would be more likely to succeed if sustainability criteria were addressed by design. For this purpose we define sustainability as 1) Having significant enrollment of the target population, 2) Having ongoing support from the major stakeholders, 3) Having financial cost that can be easily supported by the major stakeholders, and 4) Being able to engage participants while maintaining the chosen PYD framework. Program efficacy is beyond the scope of this publication.

Cyclopedia, an annual summer bicycle program, is supported by a core partnership between a medical institution (University of Rochester Medical Center, Department of Pediatrics) and a community-based organization (CBO) (Boys and Girls Club of Rochester, NY). This partnership has intrinsic properties that promote sustainability by the above criteria. First, it provides access to an existing population of at-risk youth who are members of the Boys and Girls Club (referred to herein as "the Club"). Second, it unites two key stakeholders focused on the health of children in the community, and provides broad access to other agencies and organizations with the same goal. Third, it provides access to funding sources unique to the medical field and to the community, respectively. And, fourth, it provides oversight with a network of professionals in a variety of fields who can manage the curriculum according to PYD guidelines.

Our project

The Boys and Girls Club of Rochester, NY, is located in the heart of the city's southwest quadrant on Genesee Street, which is the central north-south traffic artery through the area (see figure 2). Neighborhoods along this route include the 19th Ward, Genesee-Jefferson and Mayor's Heights. The City of Rochester, and this area in particular, differ markedly from the surrounding county. While children under 10 years old in Monroe County are 63% "White, not Hispanic," the City of Rochester is 74% "African-American and

Hispanic". Similarly, in Monroe County, 64% of children under 18 years old live in "married-couple" families, while 60% of children in the City of Rochester live in "female householder, no husband present" families. 20% of children under 12 years old live in poverty for Monroe County, but the suburbs alone are at 10% poverty compared with 43% of children in the City of Rochester (22).

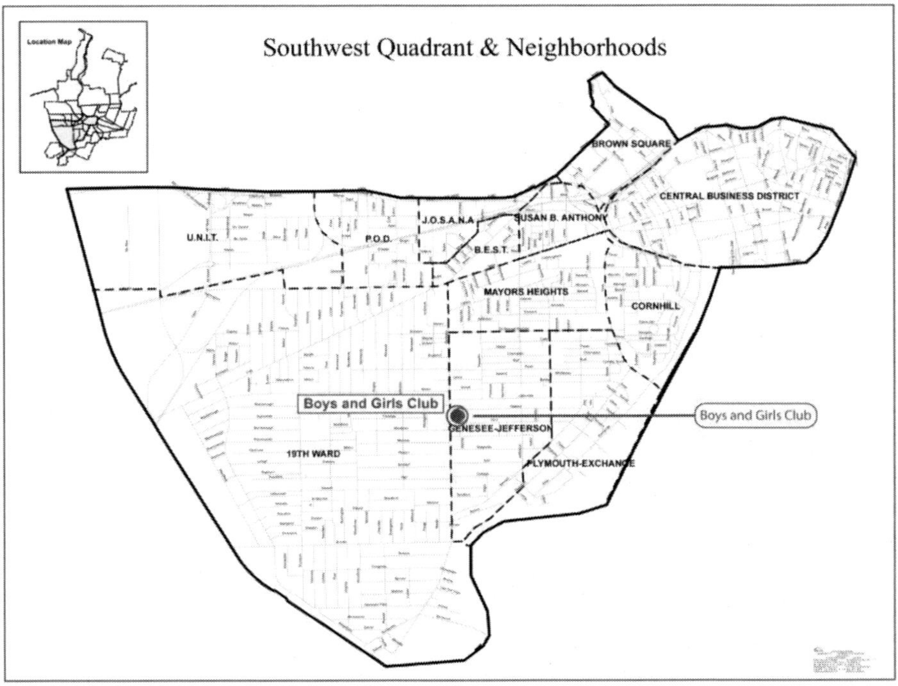

Figure 2. Rochester, NY: Southwest Quadrant. Adapted from the City of Rochester Department of Neighborhood and Business. Development Bureau of Planning & Zoning.

The Club serves children ages 6-19 years, with an annual membership of approximately 1,100 at the Genesee Street site. According to the Club's executive summary 68% of their members are male, 98% are African-American, 73% are from "single parent/skip generation households," and 91% qualify for free or reduced-price school lunches.

All Cyclopedia participants are self-selected youths (ages 12 to 17 years) who are members of the Club's Genesee Street site and who are deemed capable of riding a bicycle safely. Participants younger than 12 are directed to the "Cyclopedia Junior" program (not described here).

The Cyclopedia concept was developed under the aegis of the University of Rochester Pediatric Department's Child Advocacy Resident Education (CARE) track, a longitudinal opportunity for residents in various specialties to develop advocacy skills (mainly focusing on child advocacy), and translate those skills into community interventions over the course of their medical training. A focus on evidence-based programming and partnerships are core principles that guided Cyclopedia's endorsement of PYD programming and the search for an appropriate CBO partner (see figure 3).

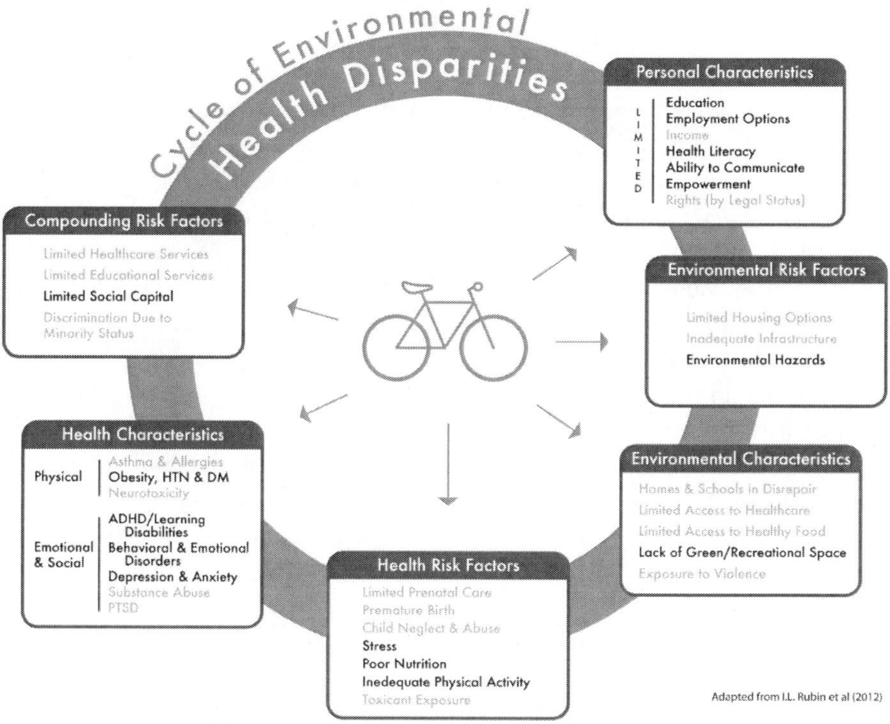

Figure 3. Cyclopedia's Direct Influence on the Cycle of Environmental Health Disparities.

The initial curriculum was developed in consultation with Club employees, following the Club's mission statement of providing "youth development programs that enable young people to acquire the skills and qualities needed to become responsible citizens and leaders." The Club has experience with PYD initiatives, and was actively seeking ways to bolster programming for its adolescent members.

The resulting program, Cyclopedia, is a bicycle program that combines physical activity with education and self-expression (writing and photography) in a 12-week curriculum of weekly or twice-weekly rides that spans the unstructured time between the end of one school year and the beginning of the next. Each ride includes a didactic lesson on a local theme (e.g. the Erie Canal, Frederick Douglass, current events), the ride relating to that theme, and work on the website where riders document their experiences through their writing and photographs.

Cyclopedia's broad goal of empowering at-risk adolescents is addressed through the component objectives: to promote physical activity; exploration; teamwork; knowledge of local geography, history and current events; communication skills; and creative self-expression.

This is delivered within a PYD framework that provides the four core constructs. The staff of the Club and the medical residents and practicing physicians who co-lead the rides are **positive adult role models**; additionally, many of the rides incorporate visits to local events and institutions where the riders are introduced to other leaders. The cumulative result is exposure to a wide variety of adult role models who participate in community organizing, education, the arts, health care, law, politics, and business.

Each ride begins with a short lecture and group discussion of the ride topic. As examples, the history of the defunct Rochester Subway introduces riders to Rochester's history as an industrial and transportation hub. The B-Boy BBQ is an annual grass-roots community event that celebrates hip-hop: graffiti art, street dancing, DJ'ing and MC'ing. The ride to visit the site or event and the riders' own writing and photography complete the **educational experience**. This is analogous to the "see one, do one, teach one" model of medical learning.

The Club is a popular institution in the neighborhood and is regarded as a haven where children can socialize, be fed, receive academic help, and participate in organized sports. Cyclopedia extends this **safe and structured environment** beyond the walls of the building with an emphasis on safe street riding and teamwork to complete each journey.

The curriculum content focuses on the riders' home city of Rochester, NY. Introduction to the people and places and events of their own city promotes a connection between each rider and his or her **sense of community**. Additionally, older riders have the opportunity to mentor younger riders and even co-lead trips.

Funding was secured through mechanisms available to each core partner. The physicians qualified for an American Academy of Pediatrics (AAP)

Community Access to Child Health (CATCH) grant and a local Community Pediatrics Advisory Council (CPAC) grant. The Club secured a Greater Rochester Health Foundation (GRHF) grant and had some discretionary funds to contribute.

The core partners also had access to several other community resources that contributed equipment and services (see table 1).

Table 1. Partner Resources

Partner (contact person)	Contribution
R Community Bikes (Dan Lill)	Bicycles (used)
Full Moon Vista Bike & Sport (Michael Spear)	Bicycle fleet maintenance
Rochester General Hospital (Gina Lord)	Bicycle helmets
Rochester Institute of Technology, College of Imaging Arts and Sciences (Oscar Palacio)	Photography printing
Individual donors (various)	Bicycles, water bottles, volunteer hours (trip-leading, curriculum development, program assessment)

On each Cyclopedia ride the names of the riders and co-leaders are recorded, along with the distance (in miles) for the trip. Riders' expressive output on the Cyclopedia website (http://www.cyclo-pedia.org) is tallied in terms of the number of written postings and, in 2011, the number of photographs.

The cost of the program is measured as the total grant money received minus the unused portions; in other words, only the spent portions factor into cost. Cost does not include donated goods or services, time donated by volunteers, or proportional share of employees' salaries.

The Mount Sinai School of Medicine Program for the Protection of Human Subjects approved this program evaluation as an "exempt" activity.

Findings

In the 2010 season, 18 male riders participated (ages 14 to 16 years). In 2011, 27 were female out of 50 total riders (ages 12 to 17 years) (see table 2).

Fifteen medical residents and physician faculty members from the University of Rochester Medical Center and seven staff members of the Club have joined rides as co-leaders. The 2010 program cumulative mileage was 743.6—this represents the sum of the miles ridden by each of the participants on the 13 offered rides. The average rider accumulated 41.3 miles. In 2011 the cumulative mileage rose to 893.2 over 20 rides; however, the average rider only accumulated 17.8 miles.

Table 2. Program Statistics

	2010	2011
Participants (% male)	18 (100)	50 (46)
Rides	13	20
Aggregate Program Miles	743.6	870
Miles/Rider	41.3	17.4
Aggregate Posts	46	27
Post/Rider	2.6	0.5
Aggregate Photos	N/A	1273
Photos/Rider		25.5

As far as writing about their experiences, riders wrote an average of 2.6 posts each in the 2010 season, and this dropped to 0.5 posts per rider in 2011.

The cost of starting the program in 2010 was $7,125, meaning the cost/rider was roughly $400. 2011 expenses dropped to $5076 for 50 riders, coming to about $100 per rider (see table 3). Additional funds provided by the GRHF and from the Club's discretionary funds were allocated to the Cyclopedia Junior program and are not included in the analysis.

Table 3. Program Cost

	Funding Source	Awarded ($)	Used ($)	Ridership	Cost/Rider ($)
2010	Resident CATCH Grant	3000	1500		
	CPAC Grant	9125	5625		
	Total	12125	7125	18	396
2011	CPAC Grant	5711	5076		
	Total	5711	5076	50	102

Conclusion

Delivery of PYD empowerment content to an at-risk adolescent population through a core partnership between a medical institution and a CBO appears sustainable as per our defined criteria, though there are deficiencies identified over the first two years of program implementation.

1. Significant enrollment of the target population

The expansion from 18 boys in the first season to 50 boys and girls in the second season suggests successful enrollment of the target population. An effort to promote the program to females for the second season resulted in 54% female composition for 2011. Compared to the Club's overall 32% female composition, this suggests Cyclopedia is a highly appealing programming option for girls. Future enrollment potential is further strengthened by the simultaneous development of a feeder program, the Cyclopedia Junior program, where approximately 30 additional children began participating in the second year.

Overall, these roughly 80 participants in 2011 represent roughly 7% of the Club's enrolled membership. As a comparison, the Club's most popular activity is the basketball program, which reaches approximately 14% of the membership. The Club offers many activities for the members, including athletic teams, leadership programs, skill-based education, and social events, and many children participate in school sports as well. Cyclopedia does not intend to replace these activities, but instead, complement them.

2. Continued support from all the stakeholders

The core partners remain committed to Cyclopedia as suggested by the number of ride co-leaders from both the medical institution and the Club. Qualitatively, support is further evinced by the CARE program's promotion of Cyclopedia as a model project for resident advocacy education. The Club uses Cyclopedia as a fundraising showpiece.

Beyond the core partnership, the aggregation of other partners has solidified Cyclopedia's sustainability and shows how the program is being embraced by multiple stakeholders. R Community Bikes, a not-for-profit bicycle repair and distribution program, donates repaired mountain bicycles to Cyclopedia, and Cyclopedia donates unusable bicycles and parts to R Community Bikes for repair and distribution to the community. Gina Lord, of the Rochester General Health System, donates free bicycle helmets each year. Michael Spear, of local bicycle shop Full Moon Vista, donates his time maintaining the bicycle fleet. Cyclopedia has participated in various community health and transportation events and organizations.

A notable deficiency is the participation of parents in the program. Each winter Cyclopedia holds an award party for all its riders, with free dinner provided. Riders who accumulate at least half the program miles receive a personalized t-shirt, wristbands representing their individual mileage achievements, and a framed print of one of their photographs. The annual party is advertised heavily to the Club community, with in-person invitations made to the majority of the parents. However, attendance from family members is low. In 2011, only two parents attended to celebrate the 2010 season, and in 2012 that number increased to six, though most were parents of the Cyclopedia Junior riders. This is unlike the parent turnout at Club basketball games and award ceremonies, which result in very high turnout from families and the community in general.

3. Financial cost that can be easily supported by the program leaders

Cost per rider continues to drop, such that monetary support for the third season (commencing May 2012) is minimal. This is largely due to the relatively large start-up cost of acquiring a functioning fleet of bicycles. The initial cost of about $400/rider for the first season dropped to about $100/rider in the second season—about 25% of the cost. Notably, a significant share of each season's budget went to the annual award party. Excluding the party expenses, the adjusted cost per rider for each season is closer to $260 for the first season and $40 for the second season, which represents about 15% of the cost.

Two important factors have made the program cost-sustainable. First, grant monies available to the medical institution combined with those available to the CBO provided ample potential sources. (Not included in the analysis is grant money to start the Cyclopedia Junior program in 2011.) Second, many costs were absorbed by key contributions of goods and services from partner organizations and individual donors, as detailed above.

4. Participant engagement while maintaining the chosen PYD framework

The Cyclopedia program was modeled on a PYD framework that provides positive adult role models, educational experiences, a safe and structured environment, and a sense of community to the participants. However, the program must appeal to the participants in order for that model to have meaning. Cyclopedia has sustained fidelity to the model, but there are signs that the curriculum is not adequately engaging the participants.

The second season saw a drop in written contributions from 2.6/rider to 0.5/rider which suggests a drop in engagement; however, this was concurrent with the introduction of cameras to the riders. The riders took 1273 photographs that season, an average of 25.5 photos/rider (see figure 4 and 5).

Figure 4. Cyclopedia riders overlooking downtown Rochester from the Pont de Rennes bridge across the Genesee River.

Figure 5. Cyclopedia riders posing with "Club" hands before a ride.

Another sign of disengagement was the drop in mileage over the two seasons, from 41.3 to 17.4 miles/rider. This essentially represents a drop in attendance, with many of the 50 riders only joining one or two rides in the second season. We suspect that older riders lost interest as younger riders joined the program. Additionally, the length of the rides, combined with unpredictable ratios of co-leaders to riders often left little time and supervision to promote the rides and provide as much attention to riders during the rides..

As Cyclopedia begins its third season in the summer of 2012, increased engagement of the riders is the priority. The ride schedule has been modified to incorporate dedicated time to writing at the end of each ride (rather than simply encouraging riders to contribute at their leisure). Each rider is now required to contribute a written entry to each ride he or she attends. Our ride schedule is posted online and weekly invitations to the community are sent through our email list server in order to promote more co-leader attendance. Parents are being invited to attend rides to encourage family investment in the program. The inclusion of younger children (12 to 14 years old) is being encouraged in order to expose each rider to more potential years of programming and to promote Cyclopedia as a part of the Club culture. The Cyclopedia Junior program continues as a "feeder program" to prime the children for bicycling as a Club activity, and continues to be popular. Older riders are encouraged to act as mentors to the younger riders in order to encourage the older riders' participation. Donation of bicycles to qualifying participants is also being considered, depending on funds.

Building on the pilot evaluation efforts in the first two years of the program, future research on the program focuses on efficacy in improving adolescent health through evaluation of metrics relating to physical fitness, obesity, school performance and social connectivity, as compared to controls.

In addition, reproducibility of the program is a goal. In 2012, the program expanded to a second location in the East Harlem neighborhood of New York City.

In summary, PYD programming has been shown to have a positive impact on reducing health disparities by preventing a variety of morbidities among at-risk youth by promoting empowerment, but those results depend on first implementing and then sustaining such programming. Cyclopedia uses a PYD framework to deliver empowerment programming along predetermined guidelines intended to promote sustainability. Continued adjustments to the curriculum have been necessary to increase participant engagement, but the program itself appears well-positioned to sustain itself by following the guidelines.

Acknowledgements

Boys and Girls Clubs of Rochester, NY, Break the Cycle, John Culhane, John Curran, Sandra-Lee Enuha, Full Moon Vista Bike & Sport (Michael Spear), Mount Sinai School of Medicine, Pediatric Environmental Health Specialty Unit, R Community Bikes (Dan Lill), Rochester General Hospital (Gina Lord), Rochester General Pediatric Associates, Rochester Institute of Technology, College of Imaging Arts and Sciences (Oscar Palacio) and University of Rochester Medical Center, Pediatric Links with the Community.

This chapter is a revised and adapted version of an earlier publication: Collins C, Lavender M, Brown S, Sheffield P, Aligne CA. Cyclopedia: Sustaining a positive youth development program through community partnership. Int J Disabil Hum Dev 2013;12(4).

References

[1] Haggerty RJ. The changing role of the pediatrician in child health care. Am J Dis Child 1974;127(4):545-9.

[2] Chamberlain LJ, Sanders LM, Takayama JI. Child advocacy training: curriculum outcomes and resident satisfaction. Ach Pediatr Adolesc Med 2005;159(9):842-7.

[3] Starfield B, Robertson J, Riley AW. Social class gradients and health in childhood. Ambul Pediatr 2002;2:238-46.

[4] Poulton R, Caspi A, Milne BJ, et al. Association between children's experience of socioeconomic disadvantage and adult health: a life-course study. Lancet 2002;360:1640-5.

[5] Marmot MG, Smith GD, Stansfeld S. et al. Health inequalities among British civil servants: the Whitehall II study. Lancet 1991;337:1387-93.

[6] Flores G, Tomany-Korman SC, Olson L. Does disadvantage start at home? Racial and ethnic disparities in health-related early childhood home routines and safety practices. Arch Pediatr Adolesc Med 2005;159(2):158-65.

[7] Smedley B, Stith A, Nelson A. eds. Unequal Treatment: Confronting Racial and Ethnic Disparities in Health Care. Washington, DC: National Academies Press, 2003.

[8] Starfield B, Shi L. The medical home, access to care, and insurance: a review of evidence. Pediatrics 2004;113(5 suppl):1493-8.

[9] Waterston T, Alperstein G, Stewart Brown S. Social capital: a key factor in child health inequalities. Arch Dis Child 2004;89:456-9.

[10] Institute of Medicine, Health literacy: A prescription to end confusion. Washington, DC: National Academies Press, 2004.

[11] Evans GW, Marcynyszyn LA. Environmental justice, cumulative environmental risk, and health among low- and middle-income children in upstate New York. Am J Public Health 2004;94:1942-4.

[12] Wright RJ, Mitchell H, Visness CM, et al. Community violence and asthma morbidity: the inner-city asthma study. Am J Public Health 2004;94(4):625-32.

[13] Singh GK, Siahpush M, Kogan MD. Rising social inequalities in US childhood obesity, 2003-07. Ann Epidemiol 2010;20:40-52.

[14] Karriker-Jaffe KJ, Foshee VA, Ennett ST. Examining how neighborhood disadvantage influences trajectories of adolescent violence: a look at social bonding and psychological distress. J Sch Health 2011;81(12):764-73.

[15] Upadhya KK, Ellen JM. Social disadvantage as a risk for first pregnancy among adolescent females in the United States. J Adolesc Health 2011;49(5):538-41.

[16] Basch CE. Teen pregnancy and the achievement gap among urban minority youth. J Sch Health 2011;81(10):614-8.

[17] Rubin IL, Geller RJ, Nodvin J, Marcus M, Howett M, Merrick J. Break the cycle of environmental health disparities in vulnerable children Int J Disabil Hum Dev 2012;11(4):00-00.

[18] Eckenrode J, Campa M, Luckey DW, Henderson CR, Cole R, Kitzman H, et al. Long-term Effects of Prenatal and Infancy Nurse Home Visitation on the Life Course of Youths19-Year Follow-up of a Randomized Trial. Arch Pediatr Adolesc Med 2010;164(1):9-15.

[19] Eccles J, Gootman JA, eds. Community programs to promote youth development. Washington, DC: National Research Council and Institute of Medicine, 2002:150-65.

[20] Schinke SP, Cole KC, Poulin SR. Enhancing the educational achievement of at-risk youth. Prev Sci 2000;1(1):51–60.

[21] Ferber T, Gaines E, Goodman C. Positive youth development: state strategies. National Conference of State Legislatures [research and policy report]. 2005 Oct 1. Accessed 2013 Jul 05. URL: http://www.ncsl.org/Portals/1/documents/cyf/final_positive _youth_development.pdf

[22] HEALTH ACTION, Priorities for Monroe County [http://www2.monroecounty.gov/health]. Rochester, NY: Monroe County Department of Public Health; [updated 2011 Feb; cited 2012 Jun 4]. Accessed 2013 Jul 05. URL: http://www2.monroecounty.gov/files/health/DataReports/ MCH%20Report%20Card,%202011%20FINAL.pdf

Section 4: Acknowledgements

In: Environmental Health
Editors: IL Rubin and J Merrick

ISBN: 978-1-62948-155-5

Chapter XIV

About the editors

I Leslie Rubin, MD, is President and Founder of the Innovative Solutions for Disadvantage and Disability (ISDD), Research Associate Professor at the Department of Pediatrics, Morehouse School of Medicine in Atlanta, Georgia, Co-director of the Southeast Pediatric Environmental Health Unit at Emory University, and Medical Director of Developmental Pediatric Specialists, Atlanta, Georgia. He is originally from South Africa, where he trained in pediatrics. He came to the United States to specialize in neonatology and then in developmental pediatrics. He was initially at the hospitals of the Case Western Reserve University in Cleveland Ohio from 1976-1980 and then moved to the Children's Hospital in Boston and the Harvard Medical School from 1980-1994. In July 1994 he moved to Atlanta, Georgia as Director of Developmental Pediatrics at Emory University and as Medical Director of the Marcus Center. Since 1998, he has been involved with the Southeast Pediatric Environmental Health Specialty Unit at Emory University, where he has integrated his understanding of developmental disabilities into the world of environmental health. In May 2004, he founded the Institute for the Study of Disadvantage and Disability (now Innovative Solutions for Disadvantage and Disability), which is dedicated to improving awareness and understanding of the relationship between social and economic disadvantage and disabilities in children. In September 2004, he left Emory University, and joined the faculty of Morehouse School of Medicine. He is the recipeint of the 2012 Calvin J Sia Award from the American Academy of Pediatrics for work on a Medical Home for Homeless Children and also through the ISDD of the 2013 Martin Lither King Jr Award from Emory University Business School and School of Public Health for the work of ISDD in improving thelives of children who grow up in circumstances of social and economic disadvantage. E-mail: lrubi01@emory.edu

Joav Merrick, MD, MMedSci, DMSc, is professor of pediatrics, child health and human development affiliated with Kentucky Children's Hospital, University of Kentucky, Lexington, Kentucky, United States and the Division of Pediatrics, Hadassah Hebrew University Medical Center, Mt Scopus Campus, Jerusalem, Israel, the medical director of the Health Services, Division for Intellectual and Developmental Disabilities, Ministry of Social Affairs and Social Services, Jerusalem, the founder and director of the National Institute of Child Health and Human Development in Israel. Numerous publications in the field of pediatrics, child health and human development, rehabilitation, intellectual disability, disability, health, welfare, abuse, advocacy, quality of life and prevention. Received the Peter Sabroe Child Award for outstanding work on behalf of Danish Children in 1985 and the International LEGO-Prize ("The Children's Nobel Prize") for an extraordinary contribution towards improvement in child welfare and well-being in 1987. E-mail: jmerrick@zahav.net.il

In: Environmental Health
Editors: IL Rubin and J Merrick
ISBN: 978-1-62948-155-5

Chapter XV

About the Innovative Solutions for Disadvantage and Disability

In 2004, the Institute for the Study of Disadvantage and Disability (ISDD) was formed to address the relationship between social and economic disadvantage and the prevalence of developmental disabilities and to reduce health disparities and the disparity in services available for this vulnerable population. In 2012 the name was changed to Innovative Solutions for Disadvantage and Disability and the mission statement was changed to more accurately reflect the work that had been done and the changing language of this field of endeavor. The mission of ISDD is to promote health equity among children living in social and economic disadvantage with or at risk for disability. ISDD accomplishes its mission thorugh three core programs:

Project GRANDD

Project GRANDD is a community-based support program for grandparents in the greater metropolitan Atlanta area, who are raising their grandchildren with disabilities, chronic illness, and behavior or learning difficulties. Project GRANDD provides group educational meetings, monthly support groups, material aid, home visits and individual case management. Project GRANDD partners with the Nell Hodgson Woodruff School of Nursing to provide students with the opportunity to provide home health visits to the families as part of their training. A significant consequence of this progam is that it has resulted in communitybuilding among the grandparents, which has promoted a sense of empowerment and independence. To date over 100 grandparents with over 200 grandchildren between them are served by Project GRANDD. This model can be applied to many popolations around the world, where grandparents have the responsibility of raising their grandchidlren – especially in areas where HIV and violent strife have decimated the adult population of parents.

Healthcare without walls

Healthcare Without Walls is a comprehensive community-based, medical home model which addresses the health care needs of children who have been homeless. ISDD works in collaboration with Mary Hall Freedom House, the Morehouse School of Medicine, Children's Healthcare of Atlanta, Emory University School of Public Health and ChildKind to provide a medical home for the children and health literacy training for their mothers.Over 100 mothers with their children havebeen served to date by thisprogram. This model is being exapanded to other populations in the Greater MetropolitanAtlanta area and is a template for improving the health and well-being of children all over the world, who suffer from a state of homelessness.

Break the cycle

This program is at the heart of ISDD in that it offers a broad look at therelationship between social and economic disadvantage and developmental disabilities and other chroninc medical conditions. ISDD works in partnership with the Southeast Pediatric Environmental Health Specialty Unit at Emory University and Break the Cycle is a multidisciplinary academic program that focuses on cultivating future leaders in the fields of medicine, nursing, public health, law, social work and education. Each year, ISDD invites students and faculty from a variety of colleges and universities to develop projects that will "Break the Cycle of Disadvantage and Disability" and raise awareness of children's environmental health disparities. The students present their projects at a national conference and the results are published in international journals. Over the past eight years Break the Cycle has engaged 71 students from 24 universities in eight states in the USA and one from Santiago Chile that have resulted in the pulication of five journal supplements. This set of projects forms the basis for this book series.

ISDD is the recipient of the 2013 Martin Luther King Jr Award from Emory University Goizueta School of Busines and the Rollins School ofPublic Health.

Contact

I Leslie Rubin, MD
President, Innovative Solutions for Disadvantage and Disability
776 Windsor Parkway
Atlanta, GA 30342 United States
E-mail: lrubi01@emory.edu
Website: www.isdd-home.org
Website: www.pehsu.emory.edu
Website: www.dpsatlanta.org

In: Environmental Health
Editors: IL Rubin and J Merrick ISBN: 978-1-62948-155-5

Chapter XVI

About the Pediatric Environmental Health Specialty Units (PEHSU)

The Pediatric Environmental Health Specialty Units (PEHSU) are an interconnected system of specialists located throughout North America (see figure 1), who respond to questions from public health professionals, clinicians, policy makers, and the public about children's environmental health concerns. The PEHSU Network provides a credible, academic, non-commercial source of health information and clinical expertise regarding environmental toxins and their effects on children.

The PEHSU Network helps people and communities with health-related decision-making; for example, advice when testing for a toxin or chemicals in air, water, and soil might be useful, andrecommendations for blood and urine testing. The most common 5 exposures of concern, based on inquiries to the National PEHSU network from 2007 to 2011, were about lead poisoning, fungus or mold, mercury, pesticides, phthalatesandBisphenol A.

There have also been responses generated around major public health concerns or such as wildfires, oil spills, carbon monoxide, chelation, and other dramatic environemntal crises like mercury spills, de-icer leaks at airports and industrial pollution in cities and towns.

While Poison Control Centers respond to acute toxic exposures, the PEHSU Network has expertise in longer term, lower dose exposures, group exposures, and what public messages and therapeutic interventions may be needed. The PEHSU Network partners closely with local, state, and federal agencies and non-profit professional organizations in carrying out this work.

The PEHSUs also provide education and consultation, and responses to a variety of environmental health topics including:

- School-related hazards
- Indoor air contamination at home
- Adolescent work place exposures
- Hazards of living near toxic waste sites
- Daycare and preschool hazards
- Inquiries about potential toxins, such as residential pesticides, childhood lead, well water contamination

The PEHSU's also provide information via the internet, with a variety of factsheets and other educational materials, educating professionals in children's environmental health through conferences, presentations and formal university based training programs and working with the media to inform the public about the facts of topics of relevance to environmental health.

This program is a critical component in the development of healthcare professionals who are better prepared to recognize, treat, and prevent toxic environmental exposures in children.

Website: www.pehsu.net

Southeast PEHSU

The Southeast PEHSU is one of the network of PEHSU's in North America that serves the southeastern states of the USA: Alabama, North Carolina, South Carolina, Florida, Georgia, Kentucky, Mississippi and Tennessee. The Southeast PEHSU is based in the Department of Pedaitrics at Emory University in Atlanta and works closely with Region 4 of the US Environmental Protection Agency (EPA) and the Agency for Toxic Substances and Disease Registry (ATSDR) at the Centers for Disease Control and Prevention (CDC). The team includes, Robert Geller, MD, pediatric toxicologist, Leslie Rubin, MD, developmental pediatrician, Maeve Howett, PhD, pediatric nurse practitioner, Janice Nodvin, BA, educator. In addition, Benjamin Gitterman, MD is a pediatrician from Washington DC who works with the Southeast PEHSU on the Break the Cycle projects and Michele Marcus, PhD, an epidemiologist who is a past member of the team and now provides consultation to the Southeast PEHSU as needed. In addition to the standard work of the PEHSU as listed above, the Southeast PEHSU works intimately with the ISDD (Innovative Solutions for Disadvantage and Disability) on the Break the CycleProgram.

Contact Southeast PEHSU

Janice Nodvin, Project Coordinator
Pediatric Environmental Health Specialty Unit
Emory University, Department of Pediatrics
49 Jesse Hill Jr Drive SE, suite 274
Atlanta, GA 30303, United States
Toll free tel: (877) 33 PEHSU or (877) 337-3478
Tel: (404) 727-9428
Fax: (770) 396-1011
E-mail: jnodvin@emory.edu
Website:www.pehsu.emory.edu

In: Environmental Health
Editors: IL Rubin and J Merrick
ISBN: 978-1-62948-155-5

Chapter XVII

About the National Institute of Child Health and Human Development in Israel

The National Institute of Child Health and Human Development (NICHD) in Israel was established in 1998 as a virtual institute under the auspicies of the Medical Director, Ministry of Social Affairs and Social Services in order to function as the research arm for the Office of the Medical Director. In 1998 the National Council for Child Health and Pediatrics, Ministry of Health and in 1999 the Director General and Deputy Director General of the Ministry of Health endorsed the establishment of the NICHD.

Mission

The mission of a National Institute for Child Health and Human Development in Israel is to provide an academic focal point for the scholarly interdisciplinary study of child life, health, public health, welfare, disability, rehabilitation, intellectual disability and related aspects of human development. This mission includes research, teaching, clinical work, information and public service activities in the field of child health and human development.

Service and academic activities

Over the years many activities became focused in the south of Israel due to collaboration with various professionals at the Faculty of Health Sciences (FOHS) at the Ben Gurion University of the Negev (BGU). Since 2000 an affiliation with the Zusman Child Development Center at the Pediatric Division of Soroka University Medical Center has resulted in collaboration around the establishment of the Down Syndrome Clinic at that center. In 2002 a full course on "Disability" was established at the Recanati School for Allied Professions in the Community, FOHS, BGU and in 2005 collaboration was started with the Primary Care Unit of the faculty and disability became part of the master of public health course on "Children and society". In the academic year 2005-2006 a one semester course on "Aging with disability" was started as part of the master of science program in gerontology in our collaboration with the Center for Multidisciplinary Research in Aging. In 2010 collaborations with the Division of Pediatrics, Hadassah Hebrew University Medical Center, Jerusalem, Israel around the National Down Syndrome Center and teaching students and residents about intellectual and developmental disabilities as part of their training at this campus.

Research activities

The affiliated staff have over the years published work from projects and research activities in this national and international collaboration. In the year 2000 the International Journal of Adolescent Medicine and Health and in 2005 the International Journal on Disability and Human Development of De Gruyter Publishing House (Berlin and New York) were affiliated with the National Institute of Child Health and Human Development. From 2008 also the International Journal of Child Health and Human Development (Nova Science, New York), the International Journal of Child and Adolescent Health (Nova Science) and the Journal of Pain Management (Nova Science) affiliated and from 2009 the International Public Health Journal (Nova Science) and Journal of Alternative Medicine Research (Nova Science). All peer-reviewed international journals.

National collaborations

Nationally the NICHD works in collaboration with the Faculty of Health Sciences, Ben Gurion University of the Negev; Department of Physical Therapy, Sackler School of Medicine, Tel Aviv University; Autism Center, Assaf HaRofeh Medical Center; National Rett and PKU Centers at Chaim Sheba Medical Center, Tel HaShomer; Department of Physiotherapy, Haifa University; Department of Education, Bar Ilan University, Ramat Gan, Faculty of Social Sciences and Health Sciences; College of Judea and Samaria in Ariel and in 2011 affiliation with Center for Pediatric Chronic Diseases and National Center for Down Syndrome, Department of Pediatrics, Hadassah Hebrew University Medical Center, Mount Scopus Campus, Jerusalem.

International collaborations

Internationally with the Department of Disability and Human Development, College of Applied Health Sciences, University of Illinois at Chicago; Strong Center for Developmental Disabilities, Golisano Children's Hospital at Strong, University of Rochester School of Medicine and Dentistry, New York; Centre on Intellectual Disabilities, University of Albany, New York; Centre for Chronic Disease Prevention and Control, Health Canada, Ottawa; Chandler Medical Center and Children's Hospital, Kentucky Children's Hospital, Section of Adolescent Medicine, University of Kentucky, Lexington; Chronic Disease Prevention and Control Research Center, Baylor College of Medicine, Houston, Texas; Division of Neuroscience, Department of Psychiatry, Columbia University, New York; Institute for the Study of Disadvantage and Disability, Atlanta; Center for Autism and Related Disorders, Department Psychiatry, Children's Hospital Boston, Boston; Department of Paediatrics, Child Health and Adolescent Medicine, Children's Hospital at Westmead, Westmead, Australia; International Centre for the Study of Occupational and Mental Health, Düsseldorf, Germany; Centre for Advanced Studies in Nursing, Department of General Practice and Primary Care, University of Aberdeen, Aberdeen, United Kingdom; Quality of Life Research Center, Copenhagen, Denmark; Nordic School of Public Health, Gottenburg, Sweden, Scandinavian Institute of Quality of Working Life, Oslo, Norway; The Department of Applied Social Sciences (APSS) of The Hong Kong Polytechnic University Hong Kong.

Targets

Our focus is on research, international collaborations, clinical work, teaching and policy in health, disability and human development and to establish the NICHD as a permanent institute at one of the residential care centers for persons with intellectual disability in Israel in order to conduct model research and together with the four university schools of public health/medicine in Israel establish a national master and doctoral program in disability and human development at the institute to secure the next generation of professionals working in this often non-prestigious/low-status field of work.

Contact

Joav Merrick, MD, MMedSci, DMSc
Professor of Pediatrics, Child Health and Human Development
Medical Director, Health Services, Division for Intellectual and Developmental Disabilities, Ministry of Social Affairs and Social Services, POB 1260, IL-91012 Jerusalem, Israel.
E-mail: jmerrick@zahav.net.il

In: Environmental Health
Editors: IL Rubin and J Merrick
ISBN: 978-1-62948-155-5

Chapter XVIII

About the book series "Pediatrics, child and adolescent health"

Pediatrics, child and adolescent health is a book series with publications from a multidisciplinary group of researchers, practitioners and clinicians for an international professional forum interested in the broad spectrum of pediatric medicine, child health, adolescent health and human development.

- Merrick J, ed. Child and adolescent health yearbook 2011. New York: Nova Science, 2012.
- Merrick J, ed. Child and adolescent health yearbook 2012. New York: Nova Science, 2012.
- Roach RR, Greydanus DE, Patel DR, Homnick DN, Merrick J, eds. Tropical pediatrics: A public health concern of international proportions. New York: Nova Science, 2012.
- Merrick J, ed. Child health and human development yearbook 2011. New York: Nova Science, 2012.
- Merrick J, ed. Child health and human development yearbook 2012. New York: Nova Science, 2012.
- Shek DTL, Sun RCF, Merrick J, eds. Developmental issues in Chinese adolescents. New York: Nova Science, 2012.
- Shek DTL, Sun RCF, Merrick J, eds. Positive youth development: Theory, research and application. New York: Nova Science, 2012.

- Zachor DA, Merrick J, eds. Understanding autism spectrum disorder: Current research aspects. New York: Nova Science, 2012.
- Ma HK, Shek DTL, Merrick J, eds. Positive youth development: A new school curriculum to tackle adolescent developmental issues. New York: Nova Science, 2012.
- Wood D, Reiss JG, Ferris ME, Edwards LR, Merrick J, eds. Transition from pediatric to adult medical care. New York: Nova Science, 2012.
- Isenberg Y. Guidelines for the healthy integration of the ill child in the educational system: Experience from Israel. New York: Nova Science, 2013.
- Shek DTL, Sun RCF, Merrick J, eds. Chinese adolescent development: Economic disadvantages, parents and intrapersonal development. New York: Nova Science, 2013.
- Shek DTL, Sun RCF, Merrick J, eds. University and college students: Health and development issues for the leaders of tomorrow. New York: Nova Science, 2013.
- Shek DTL, Sun RCF, Merrick J, eds. Adolescence and behavior issues in a Chinese context. New York: Nova Science, 2013.
- Sun J, Buys N, Merrick J. Advances in preterm infant research. New York: Nova Science, 2013.
- Tsitsika A, Janikian M, Greydanus DE, Omar HA, Merrick J, eds. Internet addiction: A public health concern in adolescence. New York: Nova Science, 2013.
- Shek, DTL, Lee TY, Merrick J, eds. Promotion of holistic development of young people in Hong Kong. New York: Nova Science, 2013.
- Shek DTL, Ma C, Lu Y, Merrick J, eds. Human developmental research: Experience from research in Hong Kong. New York: Nova Science, 2013.

Contact

Professor Joav Merrick, MD, MMedSci, DMSc
Medical Director, Medical Services
Division for Intellectual and Developmental Disabilities

Ministry of Social Affairs and Social Services
POBox 1260, IL-91012 Jerusalem, Israel
E-mail: jmerrick@zahav.net.il

Section 5: Index

Index

#

A

B

C

D

E

F

G

H

I

J

K

L

M

N

O

P

Q

R

S

T

U

V

W

Y

Z